Conducting Meta-Analysis Using SAS

Multivariate Applications Book Series

The Multivariate Applications book series was developed to encourage the use of rigorous methodology in the study of meaningful scientific issues, and to describe the applications in easy to understand language. The series is sponsored by the Society of Multivariate Experimental Psychology, and welcomes methodological applications from a variety of disciplines, such as psychology, public health, sociology, education, and business. Books can be single authored, multiple authored, or edited. The ideal book for this series would take on one of several approaches: demonstrate the application of a variety of multivariate methods to a single, major area of research; describe a methodological procedure or framework that could be applied to a variety of research areas; or present a variety of perspectives on a controversial topic of interest to applied researchers.

There are currently seven books in the series:

- *What if There Were No Significance Tests?*, co-edited by Lisa L. Harlow, Stanley A. Mulaik, and James H. Steiger (1997).

- *Structural Equation Modeling With LISREL, PRELIS, and SIMPLIS: Basis Concepts, Applications and Programming*, by Barbara M. Byrne (1998).

- *Multivariate Applications is Substance Use Research: New Methods for New Questions*, co-edited by Jennifer S. Rose, Laurie Chassin, Clark C. Presson, and Steven J. Sherman (2000).

- *Item Response Theory for Psychologists*, co-authored by Susan Embretson and Steve Reise (2000).

- *Structural Equation Modeling With AMOS: Basic Concepts, Applications, and Programming*, by Barbara M. Byrne (2001).

- *Modeling Intraindividual Variability With Repeated Measures Data: Methods and Applications*, co-edited by D. S. Moskowitz and Scott L. Hershberger (2001).

- *Conducting Meta-Analysis Using SAS*, by Winfred Arthur, Jr., Winston Bennett, Jr., and Allen I. Huffcutt (2001).

Interested persons should contact the editor, Lisa Harlow, at the University of Rhode Island, Department of Psychology, 10 Chafee Road, Suite 8, Kingston, RI 02881-0808; Phone: (401) 874-4242; Fax: (401) 874-5562; or e-mail: LHarlow@uri.edu. Information may also be obtained from one of the editorial board members: Leona Aiken (Arizona State University), Gwyneth Boodoo (Educational Testing Service), Barbara M. Byrne (University of Ottawa), Scott Maxwell (University of Notre Dame), David Rindskopf (City University of New York), or Steve West (Arizona State University).

Conducting Meta-Analysis Using SAS

Winfred Arthur, Jr.
Texas A&M University

Winston Bennett, Jr.
Air Force Research Laboratory Human Effectiveness Directorate

Allen I. Huffcutt
Bradley University

Ψ Psychology Press
Taylor & Francis Group

New York   London

First Published by Lawrence Erlbaum Associates, Inc., Publishers
10 Industrial Avenue
Mahwah, New Jersey 07430

Reprinted 2008 by Psychology Press

Cover design by Kathryn Houghtaling Lacey

Library of Congress Cataloging-in-Publication Data

Arthur, Winfred.
Conducting meta-analysis using SAS / Winfred Arthur, Jr., Winston
    Bennett, Jr., Allen I. Huffcutt.
        p. cm.
        Includes bibliographical references and index.
    ISBN 0-8058-3977-1 (cloth : alk. paper)
    ISBN 0-8058-3809-0 (pbk. : alk. paper)
    1. Psychometrics. 2. Psychology—Statistical methods. 3. Meta-
    analysis. I. Bennett, Winston. II. Huffcutt, Allen I. III. Title.
BF39 .A78 2001
150′.7′27—dc21   2001018943
                                                      CIP

Printed in the United States of America
10 9 8 7 6 5 4 3 2

CONTENTS

LIST OF TABLES

LIST OF FIGURES

Preface

FOCUS AND COVERAGE

This volume is intended to not only teach the reader about meta-analysis, but also to show the reader how to conduct one using the SAS (SAS Institute, 1990) programming code presented here. Thus, it is intended to serve as both an operational guide and a user's manual. Specifically, the volume presents and illustrates the use of the *PROC MEANS* procedure in SAS to perform the data computations called for by the two most popular and commonly used meta-analytic procedures, namely, the Hunter and Schmidt (1990a), and Glassian (Glass, McGaw, & Smith, 1981) approaches to meta-analysis. It significantly extends Huffcutt, Arthur, and Bennett's (1993) original article by including the following: the *sample-adjusted meta-analytic deviancy* (SAMD) statistic for detecting outliers; the Glassian approach to meta-analysis; the computation and interpretation of confidence and credibility intervals; and the presentation of a chi-square test for homogeneity of effect sizes. This volume also presents additional SAS programming code and examples not included in the former work and completely eliminates the manual calculation of any statistic or routine.

This volume is intended to serve as a primary resource for researchers, practitioners, instructors, and students interested in conducting a meta-analysis. Specifically, the presentation of both formulas and their associated SAS program code should keep users in touch with the technical aspects of the meta-analysis process. And, although it is assumed that the reader is reasonably familiar with the basic concepts and principles of meta-analysis, the inclusion of step-by-step instructions presented here should provide a practical means by which those unfamiliar with the meta-analysis process can not only learn, but actually conduct what is now a useful, powerful, and common approach to summarizing empirical research (Guzzo, Jackson, & Katzell, 1987; Schmidt, 1992).

In short, the volume covers both the conceptual and technical aspects of meta-analysis all within the context of using a number of SAS PROC MEANS-based programs for conducting a meta-analysis. It describes and explains the meta-analysis approach and procedures in a tutorial fashion, and then presents SAS program code that computes all the pertinent meta-analysis statistics for each approach, using examples to explain and interpret these statistics. Consequently, the volume is comprehensive and written in a tutorial tone. We do not present meta-analysis or its mathematical algorithms in excruciating detail. Our focus is more applied and practical than theoretical. In fact, one of our objectives is to fill the gap left by, and thereby complement, the theoretical books on meta-analysis (cf. Hunter & Schmidt, 1990a; Glass et al., 1981, which serve as the basis for most of the programming code presented here). Thus, this book shows and teaches the reader how to conduct and actually analyze their meta-analysis data after it has been collected.

This volume consists of five chapters. Chapter 1, presents a general overview of meta-analysis. We also present a brief review and discussion of the theory of meta-analysis, focusing on sampling error and the law of small numbers. Chapters 2 and 3 present procedures, SAS programing code, and examples of the meta-analyses of effect sizes (*d*s) and correlations (*r*s), respectively. As previously noted, our focus in these chapters is on the Glassian (Glass et al., 1981) and Hunter and Schmidt (1990a) approaches to meta-analysis. Because chapters 2 and 3 represent specific and somewhat independent approaches to meta-analysis, we made these stand-alone chapters. For instance, if the reader is only interested in the implementation of a meta-analysis of effect sizes, if the reader chooses to do so, she or he does not have to read chapter 3. Thus, chapters 2 and 3 are self-contained with limited, if any, cross-reference from one chapter to the other. This means that in those instances where specified concepts are common to both the meta-analysis of effect sizes (chap. 2) and correlations (chap. 3), they are addressed in both chapters.

Chapter 4 focuses on identifying outliers in meta-analytic data (Huffcutt & Arthur, 1995) and presents SAS code for computing the pertinent statistics for both effect sizes and correlations, along with examples and text discussing how to interpret and use these statistics. Finally, chapter 5 presents a summary of, and guidelines for, implementing a meta-analysis.

Because this book is intended to be an operational guide and user's manual, it presents multiple examples of both SAS code and output. Sample programs can also be downloaded at www.erlbaum.com. The numeric results from the hypothetical examples presented both in this volume and at the Web site have been verified by comparing them to those generated by other computer programs and software. (See Arthur, Bennett, & Huffcutt, 1994 for the specific programs including Stauffer, 1994's MetaDOS and MetaWIN programs.)

ACKNOWLEDGMENTS

To Ian and Erika who queried—"What's the story about?"—when I told them I was writing a book.

The Theory
Of Meta-Analysis —
Sampling Error and the Law
Of Small Numbers

Although not widely known, the meta-analytic technique is actually founded on a sound conceptual basis. That conceptual basis is *sampling error theory*. In psychometric terms, *sampling error* is the difference between the characteristics of a sample and those of the population from which that sample was drawn. Sampling error occurs because a sample typically represents only a small fraction of the original population. For example, if a study pertains to general human nature such as sex differences for a given personality characteristic, then a typical sample of 50 represents a meager 0.0000008 percent of the general human population of 6 billion. In such a case, any permutation of results and outcomes can and does occur, such as the effect in the sample being slightly stronger, considerably stronger, slightly weaker, or considerably weaker than the true effect in the population.

In a very real sense, sampling error is a parallel application of classical test theory. Classical test theory maintains that a person's actual score on a test is a combination of his or her true score plus an unknown (i.e., random) error component. By extension, sampling error maintains that the relationship across all of the participants in a given study, whether represented by an effect size statistic (*d*) or by a correlation coefficient (*r*), is a combination of the true size of the relationship in a population plus an unknown (i.e., random) error component.

Therein lies the logical and conceptual basis for meta-analysis. Each primary study in a meta-analysis represents one sample taken from a given population. As such, each one of those samples is likely to differ from the population by some unknown amount of sampling error. If we knew nothing at all about these sampling errors then meta-analysis would be impossible. However, we do know one key fact about sampling error that makes meta-analysis not only possible but feasible and powerful. Namely, we know that sampling errors tend to form a normal distribution with a mean of zero. By having a normal distribution with a mean of zero, it logically follows that all of the sampling errors in one direction (e.g., all studies in which the sample effect is stronger than the true population effect) will be balanced by the sampling errors in the other direction (i.e., all studies in which the sample effect is weaker than the true population effect). In short, when one computes the mean sample-weighted effect across studies in a meta-analysis, whether it be the mean effect size or the mean correlation, the resulting value is largely free of sampling error.

Consequently, the mean d value in an effect size meta-analysis is a direct estimate of what one would have obtained if it were possible to test the entire population. The true magnitude of the effect size in a population is represented by the Greek symbol "δ" (lowercase delta; see Hedges & Olkin, 1985; Hunter & Schmidt, 1990a). Accordingly, the mean effect size across the studies in an effect size meta-analysis, namely $\bar{d}$, becomes a direct estimate of δ. Similarly, the true magnitude of the correlation in a population is represented by the Greek symbol "ρ" (rho), so the mean correlation across the studies in a correlational meta-analysis, namely $\bar{r}$, becomes a direct estimate of ρ.

After estimating δ or ρ, one next assesses the variability in the aggregated ds or rs. Zero, or low levels of variability, suggest that the ds or rs represent a single population, whereas high levels of variability suggest different subpopulations (i.e., there may be moderator variables operating).

It is important to note that because meta-analytic results theoretically represent population parameters (δ and ρ), it is conceptually illogical to apply significance tests to meta-analytic results—for example, to test to see whether two ρs or δs are statistically different. First, significance tests are tests of inferences from a sample to the population; however, meta-analytic results represent population effects. Second, it has been argued that the use of significance tests in primary research under conditions of low power (almost universal in research literatures) is the very reason why research literatures appear contradictory and confusing in the first place (see Cortina & Dunlap, 1997; Hagen, 1997; Schmidt, 1992, 1996; Schmidt & Hunter, 1978; Thompson, 1996; Wilkinson et al., 1999, for a discussion of this issue), and consequently, the reason why meta-analysis is necessary to make sense of these literatures. It is, therefore, illogical to introduce this same problem to meta-analysis itself. Despite this, it is not uncommon to find meta-analytic results being subjected to inferential tests of statistical significance (see, e.g., Alliger, 1995; Aguinis & Pierce, 1998; Hedges & Olkin, 1985); and in fact, in our experience it is also not an uncommon request by journal reviewers and editors.

There are, of course, a number of issues which affect the meta-analysis process that need to be recognized. One of the most prominent of these issues is the influence of statistical artifacts such as range restriction and measurement error. These artifacts reduce the size of the effect in the primary studies, thereby making the mean effect across the studies an underestimate of the true strength of the effect in the population. Hunter and Schmidt (1990a) provide a variety of correction formulas to account for the influence of artifacts. The correction can be done individually for each study before aggregation, although that is rare because, by necessity, each study must report the appropriate artifact data. The more common approach is to correct the mean effect after aggregation using the average of whatever artifact data is available across the studies.

Another prominent issue is whether the population is homogeneous, that is, whether the strength of the effect is consistent across all situations. If an extraneous variable changes the strength of the main effect, that variable is referred to as a *moderator variable* and it causes the theory of sampling error (the fundamental basis of meta-analysis) to break down. Namely, it induces a nonrandom effect into some of the studies, thereby making the mean of the deviations from the underlying population value something other than zero. In this case the mean effect, if computed, becomes an estimate of the average strength of the effect in the population rather than an estimate of the unitary value of the effect. As noted previously, the degree of variability in the aggregated effects is one indicator of the presence or operation of moderator variables.

The standard solution when dealing with a heterogenous population is to separate the studies by the various levels of a known moderator variable and conduct a separate meta-analysis for each of the levels. If the test for homogeneity (see Hunter and Schmidt, 1990a, and in subsequent sections of this book) subsequently shows each level to be homogeneous, then each mean effect once again becomes an estimate of the unitary strength of the effect in the population (for that level of the moderator variable). If the test for homogeneity still suggests a heterogenous situation, a further breakdown by another moderator variable may be warranted.

GENERAL OVERVIEW OF META-ANALYSIS

As a general descriptive statement, meta-analysis refers to a set of statistical procedures that are used to quantitatively aggregate the results of multiple primary studies to arrive at an overall conclusion or summary across these studies. Ideally, a meta-analysis calls for several hundred data points (e.g., Arthur, Bennett, Edens, & Bell, 2001; Judge, Thoresen, Bono, & Patton, 2000; Hunter & Hunter, 1984; Schmitt, Gooding, Noe, & Kirsch, 1984). This goal is, of course, difficult to meet, although it is not uncommon for meta-analyses to sometimes consist of well over one hundred data points (e.g., Alliger, Tannenbaum, Bennett, Traver, & Shotland, 1997; Arthur, Bennett, Stanush, & McNelly, 1998; Gaugler, Rosenthal, Thornton, & Bentson, 1987; Huffcutt & Arthur, 1994; Stajkovic & Luthans, 1998). The primary reason why a less than optimal number of data points is typically analyzed is due to deficiencies in the quality of microlevel reporting (Orwin & Cordray, 1985). This occurs when primary studies fail to meet the criteria for inclusion because they do not report the information necessary to permit their inclusion in the meta-analysis. The effects and implications of a meta-analysis based on a small numbers of data points or studies are discussed in the sections on fully hierarchical moderator analysis in chapters 2 and 3.

Meta-analysis can be described as a secondary research method or design that can also be put to several uses. For instance, it can be used as a quantitative literature review procedure or a research design to test hypotheses pertaining to the relationships between specified variables. Along these lines, there are several advantages associated with meta-analysis. Meta-analyses are able to summarize large volumes of literature. They can also be used to resolve conflicts between two or more bodies of literature by comparing effect sizes across them. Conversely, they have also been known to generate their own controversies or conflicts when independent meta-analyses of ostensibly the same topic or literature have resulted in somewhat divergent conclusions. Examples of these include (a) the job performance–satisfaction relationship (Iaffaldano & Muchinsky, 1985; Judge et al., 2000; Petty, McGee, & Cavender, 1984); (b) the validity of student evaluation of instructors' teaching effectiveness (Abrami, 1984; Cohen, 1981, 1982, 1983, 1986; d'Apollonia & Abrami, 1996; d'Apollonia, Abrami, & Rosenfield, 1993; Dowell & Neal, 1982; McCallum, 1984); and (c) the personality–job performance relationship (Barrick & Mount, 1991; Hurtz & Donovan, 2000; Ones, Mount, Barrick, & Hunter, 1994; Tett, Jackson, & Rothstein, 1991; Tett, Jackson, Rothstein, & Reddon, 1994).

In fact, one can argue that meta-analysis provides a compliment to significance testing. Each method is advantageous in certain situations; having both methods available gives researchers more flexibility in the scientific research process. In particular, significance testing is advantageous when the results of an individual study must be interpreted or when the research topic is relatively new and few studies exist in that area. In contrast, meta-analysis is advantageous for research areas that are well-established and have a large number of primary studies. For example, there are hundreds of empirical studies assessing the validity of employment interviews (see Huffcutt & Arthur, 1994; Huffcutt & Roth, 1998; Huffcutt, Roth, & McDaniel, 1996; Huffcutt & Woehr, 1999; Marchese & Muchinsky, 1993; McDaniel, Whetzel, Schmidt, & Maurer, 1994; Wiesner & Cronshaw, 1988). In these areas, meta-analysis can, and has, provided information on general trends and the consistency of effects across situations.

The ability to quantitatively integrate and aggregate large numbers of studies also gives meta-analyses the ability to investigate relationships not investigated in the original primary studies (e.g., Arthur, Day, McNelly, & Edens, 2000; Arthur, Bennett, Edens, & Bell, 2001; Arthur et al., 1998; Huffcutt & Arthur, 1994), and to find trends too subtle to identify with narrative reviews (Huffcutt & Arthur, 1995). As an example, the ability to look for interactions in meta-analytic data means that quantitative reviewers can test hypotheses that were never tested in the original primary studies. Thus, a primary advantage of meta-analysis over narrative reviews is the ability to test hypotheses pertaining to the effects of moderator variables. In relation to this and in contrast to narrative reviews, meta-analysis is a more standardized and, relatively speaking, a somewhat more objective means of integrating results from multiple primary studies.

Finally, in industrial/organizational (I/O) psychology, specifically in personnel selection and test validation research, meta-analysis procedures—namely validity generalization—can be used as a validation design. For example, large-scale meta-analyses of cognitive ability tests (e.g., Hunter & Hunter, 1984; Schmidt & Hunter, 1998) have demonstrated that cognitive ability is a valid predictor of performance across a wide variety of jobs. Likewise, there are also several large-scale meta-analyses of employment interviews (e.g., Huffcutt & Arthur, 1994; Huffcutt & Woehr, 1999; Marchese & Muchinsky, 1993; McDaniel et al., 1994; Wiesner & Cronshaw, 1988) that have clearly demonstrated the moderating effect of interview structure on the criterion-related validity of employment interviews.

Although there are several forms of meta-analysis (for instance, Bangert-Drowns, 1986, identifies five of them), the most common and popular approaches are the Hunter and Schmidt (Hunter & Schmidt, 1990a; Hunter, Schmidt, & Jackson, 1982; Schmidt & Hunter, 1977), Glass (Glass, 1976; Glass et al., 1981), and Hedges and Olkin (1985) meta-analysis procedures. A distinction between these approaches is that although all three convert primary study results into a common metric (Hunter and Schmidt primarily use rs; Glass and Hedges and Olkin, ds), the Hunter and Schmidt approach—specifically referred to as validity generalization—goes a step further and corrects summary statistics for the influence of statistical artifacts such as sampling error, measurement error, and range restriction. These corrections take into account the tendency for artifacts to both reduce the average size of a relationship and increase the variability across study coefficients. Such corrections are considered to provide a more accurate estimate of the true size of a relationship (and the variance of this relationship) in the population of interest. Alternatively, Glassian meta-analysis does not typically correct for any statistical artifacts but instead, only computes and aggregates sample-weighted ds. Thus, it has sometimes been referred to as "bare-bones" meta-analysis.

Hedges and Olkin (1985) also developed their own version of meta-analysis which parallels that of Glass et al. (1981). The Hedges and Olkin approach has become popular in areas such as clinical and social psychology, and to some degree, has subsumed the Glassian approach. Although their overall goals are the same, the methodology outlined by Hedges and Olkin (1985) has some notable differences in comparison with Glass et al.'s (1981) and Hunter and Schmidt's (1990a) approaches.

At one level, meta-analytic approaches can be differentiated in terms of their focus on *correlations* (*r*s) versus *effect sizes* (*d*s). Regardless of which specific approach is used, the following 11 steps represent the general process of implementing a meta-analysis:

1. Topic selection—defining the research domain.
2. Specifying the inclusion criteria.
3. Searching for and locating relevant studies.
4. Selecting the final set of studies.
5. Extracting data and coding study characteristics.
6. Deciding to keep separate or to aggregate multiple data points (correlations or effect sizes) from the same sample—independence and nonindependence of data points.
7. Testing for and detecting outliers.
8. Data analysis—calculating mean correlations, variability, and correcting for artifacts.
9. Deciding to search for moderators.
10. Selecting and testing for potential moderators.
11. Interpreting results and making conclusions.

Although the primary focus of this volume is the data analysis step of the meta-analysis implementation process and the search and test for potential moderator variables, the next section briefly reviews the 11 steps listed previously.

Topic Selection—Defining the Research Domain

This step consists of specifying the research questions of interest. It is judgmental and is simply determined by the researcher's interest. As part of this step, the researcher may specify the independent and dependent variables and the postulated relationships between them.

Specifying the Inclusion Criteria

In this step, the researcher clearly and explicitly specifies the criteria that will be used to include, and conversely, exclude primary studies from the meta-analysis. In primary research study terms, this is analogous to specifying the research population of interest; that is, the population to which the researcher wishes to generalize her or his results. This is again a judgment call and the criteria used will obviously influence the outcomes of the meta-analysis (Abrami, Cohen, & d'Apollonia, 1988; Wanous, Sullivan, & Malinak, 1989). Thus, it is very important (indeed crucial) that the researcher clearly state what these inclusion criteria are. Examples of some inclusion criteria to consider include but are not limited to, the use of published versus unpublished studies (if only published studies are to be used, the source of these studies), the time period to be covered by the meta-analysis (e.g., 1980–2001), and the operational definitions of the variables of interest.

Searching for and Locating Relevant Studies

Searching for and locating relevant studies typically involves the use of both electronic and manual searches. The appropriateness of specified electronic databases will vary as a function of the research topic or domain. For instance, *PsychINFO* is a good database to search for behavioral research studies, but not biomedical studies. The key to a good electronic search is to identify the pertinent electronic databases and keywords or phrases to be used in the search. Electronic searches should be supplemented with manual searches which take the form of reviewing the reference lists of articles (identified in the electronic search or other seminal works), conference proceedings and programs, and contacting researchers in the topic domain to obtain additional published and unpublished studies. Searching for and locating studies is considered to be a judgmental process, so as with the other steps, the search process should be clearly documented in the meta-analysis. A listing and description of reference and information sources in the behavioral and social sciences, both electronic and print, is presented in Appendix A.

Selecting the Final Set of Studies

Selecting the final set of studies involves the application of the inclusion criteria to the studies that were located from the search. Because the application of the inclusion criteria by definition involves judgment calls, it is a good idea to have the selection of the final set of studies done independently by at least two individuals using a consensus process to resolve any disagreements.

The example presented below is from Arthur et al.'s (1998) meta-analysis of factors that influence skill decay and retention. This example is intended to illustrate the search process, the use of electronic and manual sources, and the specification and application of the inclusion criteria to arrive at the final set of studies.

Literature Search

An extensive literature search was conducted to identify empirical studies that had investigated skill retention or decay. This process started with a search of nine computer databases (*Defense Technical Information Center, Econlit, Educational Research Information Center, Government Printing Office, National Technical Information Service, PsychLit, Social Citations Index, Sociofile,* and *Wilson*). The following key words were used: skill acquisition, skill decay, skill degradation, skill deterioration, skill maintenance, skill perishability, skill retention, training effectiveness, training efficiency, and training evaluation. The electronic search was also supplemented with a manual search of the current literature. Approximately 3600 citations were obtained as a result of this initial search. A review of the abstracts of these citations for appropriate content (i.e., empirical studies that actually investigated skill retention or decay), along with a decision to retain only English language articles, narrowed the list down to 172 articles. In addition, the reference lists of these articles were reviewed and a number of researchers in the area were contacted in order to try to obtain more published and unpublished studies. As a result of these efforts, an additional 98 articles were identified, resulting in a total of 270 articles. Each article was then reviewed and considered for inclusion in the meta-analysis. The sources of these articles were as follows: journal articles (48%), technical reports (41%), books/book chapters (4%), conference papers and presentations (4%), dissertations (1%), master's thesis (1%), and unpublished or submitted manuscripts (1%).

Inclusion Criteria

A number of decision rules were used to determine the data points (studies) that would be included [in] or retained for the meta-analysis. First, to be included in the meta-analysis, a study must have investigated skill loss or retention over time with an identifiable interval of nonuse or nonpractice between the acquisition and retention test sessions. Thus, the studies had to report both preretention and postretention performance data. Second, tasks or skills were limited to "organizationally related" tasks or complex skill acquisition. Thus, for example, training interventions that focused on parent training (e.g., Therrien, 1979) were excluded. Furthermore, studies that used children as participants (e.g., Kittel, 1957; Shuell & Keppel, 1970) were also excluded. Third, to be included in the meta-analysis a study had to report sample sizes along with an outcome statistic (e.g., univariate F, t, χ^2) or other pertinent information (e.g., group means and standard deviations) that allowed the computation of or conversion to a d statistic using the appropriate conversion formulas (see Glass et al., 1981; Hunter & Schmidt, 1990a; Wolf, 1986).

Using these decision rules resulted in a retention of 52 (19%) of the 270 articles. The reasons for excluding some studies were as follows: no retention interval or not a nonpractice/nonuse retention interval (32%), insufficient statistical information to calculate or convert results to d (32%), nonempirical or not a primary study (26%), use of children/nonadult participants (4%), nonorganizational study (e.g., parent training, 3%), and unable to locate or obtain a copy of the paper (3%). (Arthur et al., 1998, pp. 67–68.)

Extracting Data and Coding Study Characteristics

This step involves reading each article in the final set and extracting data on the variables of interest, sample sizes, effect sizes, reliability of measurement, range restriction, and other specified characteristics of the study of interest to the researcher (e.g., research design, participant type, number of authors, year of publication, published vs. unpublished sources, and so on). Again, because this process is judgmental, it is a good idea to have this done independently by multiple raters. Interrater agreement should then be computed and reported. Consequently, coders should be trained to ensure that they are very familiar with the coding scheme and have a shared understanding and common frame-of-reference for the coding process and variables. Again, Arthur et al.'s (1998) documentation of this step is presented as an illustrative example.

Coding Accuracy and Interrater Agreement
The coding training process and implementation were as follows: First, the [two] coders were furnished with a copy of a coder training manual and reference guide . . . Each coder used the manual and reference guide to code a single article on their own. Next, they attended a follow-up training meeting . . . to discuss problems encountered in using the guide and the coding sheet, and to make changes to the guide, the coding sheet, or both as required. [The coders] were then assigned

TABLE 1.1
Interrater Agreement for Major Study Variables

Variable	Agreement (%)
d	95
N	90
Retention interval	95
Degree of overlearning	95
Task characteristics	
Closed-Looped vs. Open-Looped	95
Physical vs. Cognitive	100
Natural vs. Artificial	100
Speed vs. Accuracy	95
Methods of testing	100
Conditions of retrieval (similarity)	95
Evaluation criteria	
Learning criteria	100
Behavior criteria	100
OVERALL	96.67

the same five articles to code. After coding these five articles, the coders ers attended a second training session in which the degree of convergence between them was assessed. Discrepancies and disagreements related to the coding of the five articles were resolved using a consensus discussion and agreement among all four authors.

After this second meeting, the articles used in the meta-analysis were individually assigned to the coders for coding. As part of this process, the coders coded a common set of 20 articles which were used to assess the degree of interrater agreement. Interrater agreement was assessed by comparing the values recorded by each coder for each of the variables of interest. Raters were in agreement if identical values were recorded by both coders. The level of agreement obtained for the pri-

mary meta-analysis variables is presented in Table [1.1]. As these results indicate, the level of agreement is generally high with a mean
overall agreement of 96.67% ($SD = 3.12$).

Deciding to Keep Separate or to Aggregate Multiple Data Points (Effect Sizes or Correlations) From the Same Sample—Independence and Nonindependence of Data Points

The preferred practice in meta-analysis is to summarize independent data points
(effect sizes or correlations). Data points are nonindependent if they are computed
from data collected on the same group of research participants. Decisions about
nonindependence must also take into account whether the data points represent
the same variable or construct. As an example, a primary study investigating the
relationship between cognitive ability, conscientiousness, and job performance
may administer both the Wesman Personnel Classification Test (Wesman, 1965)
and the Raven Advanced Progressive Matrices (Raven, Raven, & Court, 1994;
both are measures of cognitive ability) and Goldberg's (1992) 100 Unipolar
Markers, a measure of the five-factor model of personality (and thus, conscientiousness) to the same group of participants. Such a study will then report two
ability–performance correlations, one for the Wesman and another for the Raven.
A third correlation will be reported for the conscientiousness–performance relationship. If a subsequent meta-analysis were interested in the comparative validity
of cognitive ability and conscientiousness as predictors of job performance, the
two ability correlations from this study would be considered nonindependent because they are measures of the same variable derived from the same sample. In
contrast, the ability correlations would be considered independent of the conscientiousness correlation because they are measures of different constructs. Alternatively, if the focus of the meta-analysis was the comparative validity of different operationalizations of cognitive ability, the two ability correlations would now
be considered independent because they would now be considered measures of
different variables.

For a number of reasons, *nonindependence* is an important consideration
when conducting a meta-analysis. First, one effect of nonindependence is to reduce the observed variability of the effect sizes or correlations. Under these conditions, interpretations of the homogeneity of effect sizes must be made very cautiously. Another effect of nonindependence is to artificially inflate sample sizes

beyond the number of actual participants. Although this may increase the power of the meta-analysis, it becomes difficult to determine the amount of error in the statistics describing the data points. A final effect of nonindependence is to over-weight the contribution (positively or negatively) of the studies or articles contributing multiple, nonindependent data points. Consequently, to address the problem of nonindependent data points, the accepted practice is to aggregate them by finding their average effect size or correlation.

Testing for and Detecting Outliers

A number of prominent statisticians have noted that virtually all data sets contain at least some outlier data points (Gulliksen, 1986; Mosteller & Hoaglin, 1991; Tukey, 1960, 1977). Because meta-analyses often include "studies of imperfect methodological quality, the presence of outliers is highly probable" (Schmidt et al., 1993, p. 10). Thus, an outlier in the meta-analytic framework would be a pri-mary study effect size that does not appear to be consistent with the other study effect sizes, either because of errors in the data collection or computation, unusual features of the study design, or choice of participants. Detecting outliers in meta-analytic data sets is potentially very important because the effect of such outliers is typically an increase in the residual variability and a possible shift in the mean effect size or correlation. Huffcutt and Arthur's (1995) sample-adjusted meta-an-alytic deviancy (SAMD) statistic, a technique for detecting outliers in meta-ana-lytic data sets, is presented and discussed in chapter 4.

Data Analysis—Calculating Mean Effect Sizes or Correlations, Variability, and Correcting for Artifacts; Deciding to Search for Moderators; and Select-ing and Testing for Potential Moderators

These three steps are the primary focus of this volume and are extensively dis-cussed in subsequent chapters.

Interpreting Results and Making Conclusions

Similar to the implementation of any research design or method, meta-analytic researchers must ultimately interpret their results and draw conclusions based on the interpretation of these results. The specific conclusions drawn are, of course, a function of the research questions or hypotheses of interest. However, there are several meta-analytic statistics that are commonly used to interpret the results. These statistics include the magnitude of the sample-weighted mean d or r. As a convenient guideline, Cohen (1992) describes small, medium, and large ds as 0.20, 0.50, and 0.80, respectively, and corresponding values for rs as 0.10, 0.30, and 0.50. Other statistics, extensively reviewed and discussed in chapters 2 and 3, are the variance of the effect sizes or correlations, the amount of variance accounted by sampling error and other attenuating artifacts, the magnitude of the fully corrected mean d or r, the chi-square test of homogeneity, the 75% rule, and confidence and credibility intervals.

ANALYZING META-ANALYTIC DATA AND THE ADVANTAGES OF THE SAS PROC MEANS APPROACH

With the increased popularity of meta-analysis, a number of researchers have questioned and highlighted the role of judgment calls in the implementation, and ultimately, the results of studies using this technique. One of these judgment calls obviously has to do with the data analysis step. Unlike other widely used statistical techniques such as t tests, analysis of variance, and measures of association (e.g., correlations), which are readily available in common statistical software packages such as SAS and SPSS, the researcher has to make a decision as to how the data analysis (i.e., calculating mean correlations and correcting for artifacts) will proceed when conducting a meta-analysis. These choices range from using one of several available programs without any modifications, using these programs with some modifications, writing one's own program based on the available correction formulas, to possibly doing some or all of the calculations manually. Reflecting this range of options, several programs for conducting meta-analysis on both mainframes (McDaniel, 1986ab) and personal computers (e.g., Hunter & Schmidt, 1990a; Johnson, 1989; Mullen, 1989; Stauffer, 1994) have been published.

In an investigation of the extent to which programs—all of which were based on the same conceptual and theoretical assumptions—produced identical results when used to analyze the same data set, Arthur et al. (1994) compared four commonly used Schmidt and Hunter validity generalization-based meta-analysis software programs. Their results indicated that although there were some differences in values obtained from the programs, these differences tended to be very small, typically occurred in the fourth (sometimes fifth) decimal place, and did not influence the meta-analytic outcomes or conclusions.

Consequently, because there appear to be no substantial differences in the results obtained from various meta-analysis programs, the user-friendliness, flexibility, and completeness (of the available meta-analysis statistics) of these programs becomes a very important factor. In a follow-up, Arthur, Bennett, and Huffcutt (1995) presented a detailed review of these programs and concluded that the use of SAS PROC MEANS to analyze meta-analytic data had several advantages including:

1. SAS is widely available both in personal computer and mainframe versions.
2. It is easy to use because many researchers are familiar with SAS.
3. It is highly flexible. That is, the PROC MEANS procedure is not a compiled program or software that cannot be unbundled or modified. The user can easily customize the SAS statements to correct for any number and combination of artifacts, depending on the information provided in the primary studies being analyzed. It can also be readily modified by the user to incorporate new meta-analytic advances as they appear in the extant literature.
4. Potential moderator variables can be analyzed simply by including a few additional statements in the SAS program to automatically sort and reanalyze the data according to the levels of the moderator variables.
5. The SAS PROC MEANS program also has the unique advantage of allowing the user to conduct secondary and ancillary analyses (such as correlations, regressions, t-tests, and analysis of variance) on the meta-analysis data set on the same platform without transferring data from one system or data analysis package to another. Like Hunter and Schmidt (1990a), we do not endorse the use of statistical tests to test for moderators. However, for researchers who choose to do so, the ability to run these analyses on the same platform as the meta-analysis without transferring the data to another data analysis package is particularly important.
6. The SAS PROC MEANS program also provides several meta-analysis outcome statistics.

SUMMARY

This chapter presents a general overview of meta-analysis as a secondary research method and design. A brief review of the conceptual and theoretical basis of meta-analysis highlighting the fact that it is intended to address some of the methodological problems and issues associated with small-sample primary studies and related sampling error concerns is presented. Thus, by aggregating the results of several primary studies to compute the mean effect across these studies, the resulting sample-weighted mean effect size (or correlation) is largely free of sampling error and is considered a direct estimate of the effect in the population. Other advantages of meta-analysis are also discussed.

Next, we reviewed the general steps involved in the implementation of a meta-analysis noting that the decisions associated with most of these steps are judgmental in nature and thus, have the inherent potential to influence the results of the meta-analysis. Consequently, researchers are encouraged to clearly document (or state) the decisions made in the implementation of these steps. Furthermore, wherever feasible, multiple persons should be independently involved in the implementation of these steps to permit an assessment of the degree of inter-rater agreement. Disagreements should be resolved by discussion and consensus.

Finally, the SAS PROC MEANS procedure for the analysis of meta-analytic data was introduced. The strengths and advantages of this procedure were noted. Chapter 2 describes the use of the SAS PROC MEANS procedure to run a meta-analysis of effect sizes (ds) with a focus on the Glassian approach.

Meta-Analysis
Of Effect Sizes

INTRODUCTION AND OVERVIEW

All meta-analytic approaches share one common characteristic. Namely, the results from the primary studies to be meta-analyzed must first be converted to a common metric. The two metrics used are the effect size, d, and the correlation coefficient, r. The Glassian-based meta-analytic approach (Glass, 1976; Glass et al., 1981) is probably the most popular approach to the meta-analysis of effect sizes. The Hedges and Olkin (1982) approach, a variation of the Glassian approach, has been presented as a more technically adequate form of the Glassian approach. However, Glass has restated the completeness of his original formulation (Glass & Kliegl, 1983). The meta-analysis of effect sizes is more common in research domains and disciplines that are characterized by the use of experimental or quasi-experimental research designs; certainly in those research areas that are characterized by identifiable experimental and control groups instead of correlational designs whose results are typically represented as correlation coefficients.

This chapter presents SAS code that can be used to conduct a meta-analysis of effect sizes with a primary focus on the Glassian approach. The Glassian approach involves three basic steps. First, all studies relevant to the research question or objective are collected. Next, the outcomes of each study are converted to Cohen's (1988) d statistic. After these two steps, a sample-weighted mean d and its variance are calculated and used to describe the data. The Glassian approach may also involve the test for, or analysis of, moderators. The Glassian approach does not correct for statistical artifacts and because of this, it has been described as "bare bones" (Hunter & Schmidt, 1990a) meta-analysis.

Hunter and Schmidt (1990a) present an extension of the meta-analysis of ds that corrects for sampling error and even in some instances, also corrects for criterion unreliability (Hunter & Schmidt, 1990a; Hunter et al., 1982). These extensions are presented and discussed in later sections of this chapter.

CALCULATION OF SUMMARY STATISTICS

The two summary statistics that are calculated in a Glassian meta-analysis are the sample-weighted mean d and its variance. To accomplish this, the results of all studies in the data set must first be converted to ds. The effect size, or d statistic, is the standardized difference between two means and provides a measure of the strength of a treatment or independent variable (e.g., different training methods). Thus, in experimental designs, it represents the observed difference between the experimental and the control group in standard deviation units (Cohen, 1990). A positive d value indicates that the experimental group performed better than the control group on the dependent variable. Conversely, a negative d value indicates that the control group performed better than the experimental group, and a zero d value indicates no difference between the groups. Cohen (1992) describes small, medium, and large effect sizes as 0.20, 0.50, and 0.80, respectively. Thus, a medium effect size represents half a standard deviation between means.

The d statistic is calculated as the difference between the means of the experimental (M_E) and control (M_C) groups divided by a measure of variation (Cohen, 1988; Glass, 1976; Glass et al., 1981; Hunter & Schmidt, 1990a). Glass recommends using the standard deviation of the control group as the denominator whereas Hunter and Schmidt (1990a, p. 271) use the pooled, within-group standard deviation (S_W) as the measure of variation. Thus, for posttest only experiment–control group designs where the actual means and standard deviations of the experimental and control groups are reported in the primary studies, d is computed as:

$$d = \frac{M_E - M_C}{S_W} \qquad (2.1)$$

Equations for computing S_w, when one has access to only the experimental and control group standard deviations, are presented in Appendix B.

For pretest–posttest experimental–control group designs, d can be computed as:

$$d = \frac{(M_{ET2} - M_{ET1}) - (M_{CT2} - M_{CT1})}{S_{ET1T2,CT1T2}} \qquad (2.2)$$

where M_{ET1} and M_{ET2} are the pretest and posttest means of the experimental group, M_{CT1} and M_{CT2} are the pretest and posttest means of the control group, and $S_{ET1T2,CT1T2}$ is the pooled pretest and posttest standard deviations of the experimental and control groups.

For single group pretest–posttest designs, d can be computed as:

$$d = \frac{M_{T2} - M_{T1}}{S_{T1T2}} \qquad (2.3)$$

where M_{T2} is the posttest mean, M_{T1} is the pretest mean, and S_{T1T2} is the pooled pretest and posttest standard deviation (cf. Carlson & Schmidt, 1999, who use only the pretest standard deviations as the denominator for Equations 2.2 and 2.3). Single group ds should be corrected for upward biasing effects using Dunlap, Cortina, Vaslow, & Burke's (1996) procedures (see Appendix C) for computing effect sizes from matched, repeated measures, or single group designs.

So, for studies that report actual means and standard deviations of the experimental and control groups, effect sizes can be directly calculated by inserting these values into Equations 2.1–2.3. For studies that report other outcome statistics (e.g., correlations, t statistics, or univariate two-group F statistics), conversion formulas appropriate to the specified statistic must be used to compute ds. Examples of some commonly used conversion formulas are provided in Appendix C.

Rosnow, Rosenthal, and Rubin (2000) also present a number of formulas and procedures for computing standardized measures of effect sizes.

In calculating the mean overall d (i.e., the average d across all studies) and its variance, which are denoted as $\bar{d}$ and $Var(d)$, respectively, the individual study ds are sample weighted. This results in assigning more weight to studies with larger sample sizes compared to those with smaller samples. The conceptual basis for sample-weighting is that larger sample studies have less sample error than smaller sample studies and thus, should be assigned more weight in the computation of the mean effect. As such, the mean effect becomes a reasonable estimate of the true strength of the effect in the population. The equations for calculating these two summary statistics are similar to those presented in Equations 3.1 and 3.2 for the meta-analysis of correlations (see chap. 3). Specifically, these equations are (Hunter & Schmidt, 1990a, p. 285):

$$\bar{d} = \frac{\Sigma(N_i - d_i)}{\Sigma N_i} \tag{2.4}$$

$$Var(d) = \frac{\Sigma(N_i * (d_i - \bar{d})^2)}{\Sigma N_i} \tag{2.5}$$

where N_i is the sample size for each study and d_i is the observed effect size. The SAS statements necessary to calculate the sample-size-weighted mean d and the sample-size-weighted variance are presented in Table 2.1. The reader should note that all the SAS code and associated examples presented in this volume were ran on a mainframe. A hypothetical meta-analysis of ten study ds is presented in order to illustrate the procedures and to allow readers the chance to practice with known values. These ten studies can be considered to be evaluation studies that assessed the effectiveness of a specified type of training program. Five of the studies used objective criteria, and the other five used subjective criteria. The results of these analyses, along with the line number of program codes producing specified results, are presented in Table 2.4.

TABLE 2.1
SAS Statements for Conducting a Meta-Analysis of Effect Sizes
With Correction for Sampling Error

```
001   DATA OBSERVED;            ****************** COMMENTS ************************;
002   INPUT STUDY D N CRITERIA $;***   STUDY = STUDY NUMBER, D = EFFECT SIZE    ********;
003   CARDS;                    ***   N = CORRESPONDING SAMPLE SIZE             ********;
004    1    .34    100    OBJ
005    2    .30    40     SBJ
006    3    .41    65     SBJ
007    4    .70    75     SBJ
008    5    .54    85     OBJ
009    6    .87    50     OBJ
010    7    .32    20     OBJ
011    8    .77    25     SBJ
012    9    .29    35     SBJ
013   10    .47    90     OBJ
014   ;
015   DATA SMP_WGT1;
016      SET OBSERVED;
017   PROC MEANS MEAN VAR STD MIN MAX RANGE VARDEF=WEIGHT;
018      VAR D;
019      WEIGHT N;                          *************************************************;
020      OUTPUT OUT=ADATA MEAN=SWMD VAR=VAR_D; * SWMD=SAMPLE-WEIGHTED MEAN D * *;
021   DATA SMP_WGT2;                        ***    MEAN_N = AVERAGE N ACROSS        ***;
022      SET OBSERVED;                      ***    ALL STUDIES, N=TOTAL N           ***;
023   PROC MEANS MEAN SUM N MIN MAX RANGE; ***    K=NUMBER OF STUDIES             ***;
024      VAR N;                             *************************************************;
025      OUTPUT OUT=BDATA MEAN=MEAN_N SUM=N N=K;
```

Note. The input variable D is the d (effect size) for each study and the variable N is the corresponding sample size. Most SAS users read in their data from an external data file. However, for the sake of clarity of presentation, we have chosen to present and use the data as part of the SAS program file. Finally, depending on the computer system, it may be necessary to place certain job control language (JCL) statements before the SAS statements presented in Table 2.1.

The first *PROC MEANS* (program code line 017) calculates the sample-size-weighted mean and the sample-size-weighted variance for the ds. The second *PROC MEANS* (program code line 023) calculates the average sample size across the ten studies (*MEAN_N*), the overall/total sample size across the ten studies (N), and total number of studies/data points (k). Running this program with the hypothetical data results in a sample-weighted mean d of 0.5002564 (min = 0.29, max = 0.87, range = 0.58) and a variance across ds of 0.0320726, both sample-size-weighted, and an average sample size of 58.50 (min = 20, max = 100, range = 80).

TABLE 2.2
SAS Statements for Computing Confidence Intervals, a Chi-Square Test
for the Homogeneity of *d*s, and Other Test Statistics
for the Meta-Analysis of Effect Sizes

```
026    DATA FINALE;
027      MERGE ADATA BDATA;
028        VAR_E = ((MEAN_N - 1) / (MEAN_N - 3)) * (4 / MEAN_N) * (1 + ((SWMD ** 2) / 8));
029        VAR_DLTA = VAR_D - VAR_E;
030    IF VAR_DLTA LT 0 THEN VAR_DLTA = 0;
031        PVA_SE = 100 * (VAR_E / VAR_D);
032        SD_DLTA = SQRT (VAR_DLTA);
033        L95_CONF = SWMD - (1.96 * SD_DLTA);
034        U95_CONF = SWMD + (1.96 * SD_DLTA);
035        CHI_SQ = K * VAR_D / VAR_E;
036    PROC PRINT;
037        VAR SWMD VAR_DLTA SD_DLTA VAR_E PVA_SE
038            L95_CONF U95_CONF CHI_SQ;
```

Note. In SAS, underscores must be used in variable names rather than a hyphen (e.g., *VAR_D* vs. *VAR-D)* to avoid confusion with the mathematical operator for subtraction.

SWMD is the sample-weighted mean *d*. *VAR_D* is the sample-size-weighted variance. *VAR_E* is the variance attributable to sampling error. *VAR_DLTA* and *SD_DLTA* are the variance and standard deviation of the population *d*, respectively. *L95_CONF* and *U95_CONF* are the lower and upper bounds of the 95% confidence interval. The 1.96 should be replaced with the appropriate *z* value for the specified interval (i.e., 99% = 2.58; 90% = 1.64). *CHI_SQ* is the calculated chi-square test statistic ($df = k$-1) of homogeneity of the *d*s (i.e., is the unexplained variance significantly greater than zero?). The user must refer to a chi-square table to determine the significance of this test at the specified *p* value. A chi-square table is presented in Appendix D.

Note that as Cohen (1990, 1994) has quite eloquently discussed, it is rarely necessary to present results past two decimal places. However, to not confuse the reader, we have chosen to report the exact numbers obtained from the SAS printout (see Table 2.6). This is obviously with the clear understanding that in reporting the results in a research paper or technical report, one would typically round off most of the meta-analytic results to two decimal places (e.g., Tables 2.7, 2.10, and 2.13).

ESTIMATION OF SAMPLING ERROR VARIANCE

As noted previously, the basic Glassian meta-analysis of *d*s does not correct for statistical artifacts but simply computes only the sample-weighted mean effect size. Hunter and Schmidt (1990a) have presented extensions of the meta-analysis of ds which correct for sampling error and even criterion unreliability. We next present these extensions to the meta-analysis of ds.

Hunter and Schmidt (1990a) noted that, in most cases, the largest source of the variability across study effect sizes is sampling error. The effect of sampling error is that study ds will vary randomly from the population value (δ). The removal of the variability across study ds that can be attributed to sampling error results in a more accurate estimate of the true variance across ds. The standard formula for sampling error variance of d values is used to determine how much variance would be expected in the observed ds from sampling error alone. This is calculated using Equation 2.6, where $\bar{d}$ is the sample-size-weighted mean d calculated according to Equation 2.4, and $\bar{N}$ is the average sample size. Equation 2.6 (Hunter & Schmidt, 1990a, pp. 286; 289) is presented below along with the corresponding calculations for the hypothetical example.

$$Var(e) = \left[\frac{\bar{N}-1}{\bar{N}-3}\right]*\left[\frac{4}{\bar{N}}\right]*\left[1*\left(\frac{\bar{d}^2}{8}\right)\right] \qquad (2.6)$$

$$Var(e) = \left[\frac{58.50-1}{58.50-3}\right]*\left[\frac{4}{58.50}\right]*\left[1*\left(\frac{0.50^2}{8}\right)\right] \qquad (2.6.1)$$

$$Var(e) = 0.073056 \qquad (2.6.2)$$

In Table 2.2, program code lines 028 and 036-038 produce the same result. Removing the variance attributable to sampling error from the total variance across ds is a simple matter of subtracting $Var(e)$ from the total variance across ds, $Var(d)$. Thus, the variance of the population effect sizes, $Var(\delta)$, is estimated as:

$$Var(\delta) = Var(d) - Var(e) \tag{2.7}$$

$$Var(\delta) = 0.0320726 - 0.073056 \tag{2.7.1}$$

$$Var(\delta) = -0.040984 \tag{2.7.2}$$

$Var(\delta)$ is presented in the SAS program code as *VAR_DLTA*, lines 029 and 036-038 in Table 2.2. Program code line 030 sets *VAR_DLTA* to zero (0) if it is less than zero (i.e., if it is a negative value), as is the case in our specific example. This is necessary in order to compute the standard deviation of $Var(\delta)$; (which is the square-root of $Var(\delta)$; see Equation 2.8). That is, if $Var(\delta)$ is negative, its square-root will be an invalid mathematical operation.

Thus, in the absence of any moderators, the best estimate of the effectiveness of the training program in the hypothetical example is an average sample-weighted effect size of 0.50 with a standard deviation of 0.

As previously indicated, the d statistic is a standard deviation metric used to express the difference between treatment and control groups, usually in experimental studies; that is, it represents the standardized difference between group means. In situations where the sample sizes of the groups are very uneven, Hunter and Schmidt (1990a) recommend "correcting" the mean sample-weighted d ($\bar{d}$) for the attenuating effect of unequal or unbalanced sample sizes. This is accomplished using a bias multiplier, denoted as A, which is calculated as (Hunter & Schmidt, 1990a, pp. 281–283; 289):

$$A = 1 + \left[\frac{0.75}{\overline{N} - 3} \right] \tag{2.8}$$

It should be noted that for sample sizes of 100 or larger, the bias multiplier will differ only trivially from 1.00. In our example,

$$A = 1 + \left[\frac{0.75}{58.50 - 3} \right] \tag{2.8.1}$$

$$A = 1.03151 \tag{2.8.2}$$

In Table 2.3, program code lines 031.1 and 036-038 produce the same result. The corrected mean d and standard deviation of the population effect sizes (SD_δ) are then obtained by dividing the mean d and standard deviation by the bias multiplier. It should be noted that to the extent that A is larger than 1.00, dividing the mean d by A will result in a corrected mean d that is smaller than the uncorrected mean d. In our example, the corrected mean d ($\overline{d}_c$) and standard deviation (SD_δ) are calculated as follows:

$$\overline{d}_c = \frac{\overline{d}}{A} \tag{2.9}$$

$$\overline{d}_c = \frac{0.5002564}{1.03151} \tag{2.9.1}$$

$$\overline{d}_c = 0.49359 \tag{2.9.2}$$

and

$$SD_\delta = \frac{\sqrt{Var(\delta)}}{A} \tag{2.10}$$

TABLE 2.3
SAS Statements for Correcting Mean Sample-Weighted _d_ for the
Effect of Unequal Sample Sizes

```
026      DATA FINALE;
027        MERGE ADATA BDATA;
028          VAR_E = ((MEAN_N - 1) / (MEAN_N - 3)) * (4 / MEAN_N) * (1 + ((SWMD ** 2) / 8));
029          VAR_DLTA = VAR_D - VAR_E;
030      IF VAR_DLTA LT 0 THEN VAR_DLTA = 0;
031          PVA_SE = 100 * (VAR_E / VAR_D);
031.1        A = 1 + (.75 / (MEAN_N - 3));
032          SD_DLTA = SQRT (VAR_DLTA);
032.1        SWMD_C = SWMD / A;
032.2        SD_DLT_C = SD_DLTA / A;
033          L95_CONF = SWMD_C - (1.96 * SD_DLT_C);
034          U95_CONF = SWMD_C + (1.96 * SD_DLT_C);
035          CHI_SQ = K * VAR_D / VAR_E;
036      PROC PRINT;
037          VAR SWMD_C VAR_DLTA SD_DLTA SD_DLT_C VAR_E PVA_SE
038              L95_CONF U95_CONF CHI_SQ A;
```

Note. Table 2.3 is a revision of Table 2.2 with the addition of SAS code (031.1, 032.1, and 032.2) to correct for the effect of unequal or unbalanced sample sizes. The variable name _SWMD_ in lines 033, 034, and 037 has been changed to _SWMD_C_. Also, in lines 033 and 034, _SD_DLTA_ has been changed to _SD_DLT_C_.

$$SD_\delta = \frac{\sqrt{0.00}}{1.03151} \qquad\qquad (2.10.1)$$

$$SD_\delta = 0.00 \qquad\qquad (2.10.2)$$

where _Var(δ)_ is the population variance.

　　　The specific SAS code for Equations 2.8 to 2.10.2 are not presented in Tables 2.1 and 2.2. Instead, the SAS code for correcting the mean sample-weighted _d_ for the effect of unequal or unbalanced sample sizes is presented in Table 2.3. Specifically, Table 2.3 is a revision of Table 2.2 with the addition of SAS code to correct for unequal sample sizes. $\bar{d}_c$ is denoted as _SWMD_C_ in the SAS program code lines 032.1, 033, 034, and 036-038 in Table 2.3 and SD_δ is denoted as SD_DLT_C in the SAS program code lines 032.2, 033, 034, and 036-038 in Table 2.3.

Finally, as with the Hunter and Schmidt (1990a) approach to the meta-analysis of correlations, the degree to which sampling error increases variability can be expressed by computing the percent of the observed variance that is accounted for by sampling error. This calculation is shown in Equation 2.11 for the hypothetical example, where the percent of variance accounted for by sampling error is denoted as PVA_{SE}.

$$PVA_{SE} = \left[\frac{Var(e)}{Var(d)}\right]*100 \tag{2.11}$$

$$PVA_{SE} = \left[\frac{0.073056}{0.0320726}\right]*100 \tag{2.11.1}$$

$$PVA_{SE} = 227.784 \tag{2.11.2}$$

This result is also produced by program code lines 031 and 036-038 in Table 2.2. On occasion sampling error variance will actually be larger than the observed variance across the effect sizes from the studies, causing more than 100% of the observed variance to be accounted for by sampling error, as was the case in our example. A key point is that the data set for a particular meta-analysis represents only one possible set of studies taken from the target population. As such, the variability across these studies does not always represent the true variability in the population exactly. When it underestimates the true population variability, the possibility exists that the estimate of sampling error variance will exceed the observed variance. When this happens, the logical course of action and standard practice is to substitute 100% for the actual percent-of-variance estimate. Doing so provides for a clearer presentation of the results and does not change the basic premise that no moderator variables are operating.

CONFIDENCE INTERVALS

Confidence intervals are used to assess the accuracy of the estimate of the mean effect size or correlation (i.e., $\bar{d}$ or $\bar{r}$; Whitener, 1990). Confidence intervals estimate the extent to which sampling error remains in the sample-size-weighted mean effect size. A confidence interval gives the range of values that the mean effect size is likely to take if other sets of studies were taken from the population and used in the meta-analysis. A 95% confidence interval for the mean d is calculated as (Hunter & Schmidt, 1990a, pp. 283; 288):

$$Conf_{95\%} = \bar{d} \pm (1.96) * (SD_\delta) \qquad (2.12)$$

which in our specific example is:

$$Conf_{95\%} = 0.50026 \pm (1.96) * (0) \qquad (2.12.1)$$

$$Conf_{95\%} = 0.50026 \pm 0 \qquad (2.12.2)$$

Thus, the 95% confidence interval for our corrected mean d is:

$$0.50026 < \bar{d} < 0.50026 \qquad (2.12.3)$$

These results are produced by program code lines 033-034 and 036-038 in Table 2.2. [Note: The 1.96 should be replaced with the appropriate z value for the specified confidence interval (i.e., 99% = 2.58; 90% = 1.64)].

TABLE 2.4
Meta-Analysis of Effect Sizes Results
for Hypothetical Training Effectiveness Data

Meta-Analysis Statistic	Results	Program Code Line Numbers From Tables 2.1 and 2.2
Sampling Error Results		
Total N	585	023-024
Mean N	58.50	023-024
Min N	20	023-024
Max N	100	023-024
Range of N	80	023-024
Number of ds	10	023-024
Mean observed d (sample-weighted)	0.5002564	017-019
Min observed d	0.29	017-019
Max observed d	0.87	017-019
Range of observed ds	0.58	017-019
Variance of observed ds	0.0320726	017-019
SD of observed ds	0.1790882	017-019
Variance due to sampling error	0.073056	028; 036-038
Percent variance accounted for	227.784[*]	031; 036-038
Variance δ	0.00	029; 036-038
SD_δ	0.00	032; 036-038

(continued)

95% Confidence Interval

Lower limit	0.50026	033; 036-038
Upper limit	0.50026	034; 036-038
Chi-square value	4.39013	035; 036-038

Note. *On occasion, sampling error variance will actually be larger than the observed variance across the effect sizes from the studies, causing more than 100% of the observed variance to be accounted for by sampling error, as was the case in our example. A key point is that the data set for a particular meta-analysis represents only one possible set of studies taken from the target population. As such, the variability across these studies does not always exactly represent the true variability in the population. When it underestimates the true population variability, the possibility exists that the estimate of sampling error variance will exceed the observed variance. When this happens, the logical course of action and standard practice is to substitute 100% for the actual percent-of-variance estimate. Doing so provides for a clearer presentation of the results and does not change the basic premise that no moderator variables are operating.

The SAS program code and the results of the hypothetical example are presented in Tables 2.1, 2.2, and 2.4, respectively.

CORRECTING SUMMARY STATISTICS FOR ARTIFACTS—CRITERION RELIABILITY

As discussed more extensively later in chapter 3, criterion unreliability attenuates correlations; this effect is also true for ds. That is, criterion unreliability causes rs and ds to be suppressed from their true values. Consequently, in situations where the measure used to operationalize the dependent variable (i.e., criterion) has less than perfect reliability, it has been suggested (e.g., Hunter & Schmidt, 1990a; Hunter et al., 1982) that the mean d must be corrected for the attenuating effect of this measurement error in the same way as the correlation coefficient is corrected in the Hunter and Schmidt (1990a) approach to the meta-analysis of correlations. Thus, the corrected $\bar{d}$ ($\bar{d}_T$) is calculated as (Hunter et al., 1982, p. 111):

$$\overline{d}_T = \frac{\overline{d}}{\sqrt{r_{xx}}} \qquad (2.13)$$

where $\overline{d}$ is the sample-weighted d, and r_{xx} represents the criterion's reliability.

Although the specific SAS code for Equation 2.13 is not presented in Tables 2.1 or 2.2, it can be easily written and included in the SAS program code. As an illustrative example, take a self-efficacy training study that uses an experimental design [i.e., has both an experimental (training) group, and a control (no training) group] but uses a less than perfect measure (reliability of 0.60) to operationalize the dependent variable, self-efficacy. The solution for Equation 2.13 for this example will be:

$$\overline{d}_T = \frac{0.5002564}{\sqrt{0.60}} \qquad (2.13.1)$$

$$\overline{d}_T = 0.6458282 \qquad (2.13.2)$$

The SAS code for Equation 2.13 for this example will be:

$$SWMD\_T = SWMD / SQRT \, (0.60);$$

In the examples presented in Tables 2.1 and 2.2, this line of code will be inserted after program code line 032. And if we do so, the variable name $SWMD$ in lines 033 and 034 must be changed to $SWMD\_T$. This example of SAS code for correcting for criterion unreliability is presented Table 2.5. Specifically, Table 2.5 is a revision of Table 2.2 with the inclusion of SAS code (program code line 032.1) to correct for criterion unreliability (for a dependent variable measure with a reliability of 0.60).

TABLE 2.5
**SAS Statements for Correcting for Criterion Unreliability
(and Computing Confidence Intervals and a Chi-Square Test for the
Homogeneity of _d_s) for the Meta-Analysis of Effect Sizes**

```
026      DATA FINALE;
027        MERGE ADATA BDATA;
028          VAR_E = ((MEAN_N - 1) / (MEAN_N - 3)) * (4 / MEAN_N) * (1 + ((SWMD ** 2) / 8));
029          VAR_DLTA = VAR_D - VAR_E;
030      IF VAR_DLTA LT 0 THEN VAR_DLTA = 0;
031          PVA_SE = 100 * (VAR_E / VAR_D);
032          SD_DLTA = SQRT (VAR_DLTA);
032.1        SWMD_T = SWMD / SQRT (.60);
033          L95_CONF = SWMD_T - (1.96 * SD_DLTA);
034          U95_CONF = SWMD_T + (1.96 * SD_DLTA);
035          CHI_SQ = K * VAR_D / VAR_E;
036      PROC PRINT;
037          VAR SWMD_T VAR_DLTA SD_DLTA VAR_E PVA_SE
038                  L95_CONF U95_CONF CHI_SQ;
```

Note. Table 2.5 is a revision of Table 2.2 with the addition of SAS code (program line 032.1) to correct for criterion unreliability (for a dependent variable measure with a reliability of .60).

CHI-SQUARE TEST FOR HOMOGENEITY

A chi-square test for homogeneity can be used to assess and test for the effects of moderator variables. This test assesses whether the residual variance is significantly different from zero. As with the meta-analysis of correlations, specifically the Hunter and Schmidt (1999a) approach (see chap. 3 of this volume), if moderator variables are operating, there may be more than one population involved and possible subpopulations should be investigated.

However, it should be noted that Hunter and Schmidt (1990a) "do not endorse this significance test because it asks the wrong question. Significant variation may be trivial in magnitude, and even nontrivial variation may still be due to research artifacts" (p. 110). They also go on to caution that "if the meta-analysis has very many studies [data points], it has very high statistical power and will therefore reject the null hypothesis, given even a trivial amount of variation across studies" (p. 112).

A notable difference between the use of the chi-square test in the meta-analysis of correlations, specifically, the Hunter and Schmidt (1990a) approach and its

use in the meta-analysis of effect sizes (i.e., the Glassian approach) is that, because the Glassian approach does not correct for any attenuating artifacts (and
thus, their label "bare bones" meta-analysis), the test for homogeneity (extent of
sampling error) is determined by computing a comparison variance ratio multi-
plied by the number of data points. This test is thus, based on the sample-size-
weighted ds. The test statistic, Q, will have a chi-square distribution with k-1 degrees of freedom and using Hunter and Schmidt's formula (1990a, p. 428) is calculated as:

$$Q_{(k-1)} = k*\left[\frac{Var(d)}{Var(e)}\right] \tag{2.14}$$

$$Q_{(10-1)} = 10*\left[\frac{0.032076}{0.073056}\right] \tag{2.14.1}$$

$$Q_{(9)} = 4.39013 \tag{2.14.2}$$

where k is the number of ds (data points), $Var(d)$ is the variance of the sample-
weighted ds (Equation 2.5; program code lines 017-019 in Table 2.1), and $Var(e)$
is the sampling error variance (Equation 2.6; program code lines 028 and 036-038
in Table 2.2). Equation 2.14 is also presented in the SAS program code in Table
2.2 (lines 035-038). Our chi-square value is not significant at $p = .05$ (the tabled
value required for significance is 16.92). The results of the hypothetical example
are presented in Table 2.4.

SAS PRINTOUT FOR TRAINING EFFECTIVENESS EXAMPLE

The SAS printout for the training effectiveness example used in Tables 2.1, 2.2,
and 2.4, is presented in Table 2.6. This printout includes both the SAS program
code as well as the output (i.e., results).

TABLE 2.6
SAS Printout for the Meta-Analysis of Effect Sizes
for Hypothetical Training Effectiveness Data

```
1                              The SAS System        22:34 Sunday, March 19, 2000

NOTE: Copyright (c) 1989-1996 by SAS Institute Inc., Cary, NC, USA.
NOTE: SAS (r) Proprietary Software Release 6.09  TS470

*****************************************************************
*                                                               *
*        Welcome to the new SAS System, Release 6.09 Enhanced   *
*                                                               *
*****************************************************************

NOTE: SAS system options specified are:
      SORT=4

NOTE: The initialization phase used 0.17 CPU seconds and 2041K.
1          OPTIONS LS=90;
2          DATA OBSERVED;              ***        COMMENTS           **;
3           INPUT STUDY D  N  CRITERIA $; *** STUDY=STUDY NUMBER, D=EFFECT SIZE **;
4           CARDS;

NOTE: The data set WORK.OBSERVED has 10 observations and 4 variables.
NOTE: The DATA statement used 0.05 CPU seconds and 2679K.

4                              *** N = CORRESPONDING SAMPLE SIZE    **;
15          ;
16         DATA SMP_WGT1;
17          SET OBSERVED;

NOTE: The data set WORK.SMP_WGT1 has 10 observations and 4 variables.
NOTE: The DATA statement used 0.01 CPU seconds and 2683K.

18         PROC MEANS MEAN VAR STD MIN MAX RANGE VARDEF=WEIGHT;
19          VAR D;
20          WEIGHT N;
21          OUTPUT OUT=ADATA MEAN=SWMD  VAR=VAR_D; ****************************;
22                         ** MEAN_N=AVERAGE N ACROSS  **;

NOTE: The data set WORK.ADATA has 1 observations and 4 variables.
NOTE: The PROCEDURE MEANS printed page 1.
NOTE: The PROCEDURE MEANS used 0.03 CPU seconds and 2888K.

23         DATA SMP_WGT2;              ** ALL STUDIES, N=TOTAL N   **;
24          SET OBSERVED;             ** K=NUMBER OF STUDIES       **;

NOTE: The data set WORK.SMP_WGT2 has 10 observations and 4 variables.
NOTE: The DATA statement used 0.01 CPU seconds and 2888K.

2                              The SAS System        22:34 Sunday, March 19, 2000

25         PROC MEANS MEAN SUM N MIN MAX RANGE;   *****************************;
26          VAR N;
27          OUTPUT OUT=BDATA MEAN=MEAN_N  SUM=N  N=K;
28

NOTE: The data set WORK.BDATA has 1 observations and 5 variables.
NOTE: The PROCEDURE MEANS printed page 2.
NOTE: The PROCEDURE MEANS used 0.01 CPU seconds and 2888K.

29         DATA FINALE;
30          MERGE  ADATA BDATA;
31           VAR_E = ((MEAN_N - 1) / (MEAN_N - 3)) * (4 / MEAN_N) *
32              (1 + ((SWMD ** 2) / 8));
33           VAR_DLTA = VAR_D - VAR_E;
34          IF VAR_DLTA LT 0 THEN VAR_DLTA = 0;
35           PVA_SE = 100 * (VAR_E / VAR_D);
36           SD_DLTA = SQRT (VAR_DLTA);
```

(continued)

```
37              L95_CONF = SWMD - (1.96 * SD_DLTA);
38              U95_CONF = SWMD + (1.96 * SD_DLTA);
39              CHI_SQ = K * VAR_D / VAR_E;

NOTE: The data set WORK.FINALE has 1 observations and 17 variables.
NOTE: The DATA statement used 0.04 CPU seconds and 3005K.

40         PROC PRINT;
41            VAR  SWMD  VAR_DLTA  SD_DLTA  VAR_E  PVA_SE
42              L95_CONF  U95_CONF  CHI_SQ;

NOTE: The PROCEDURE PRINT printed page 3.
NOTE: The PROCEDURE PRINT used 0.02 CPU seconds and 3088K.

NOTE: The SAS session used 0.36 CPU seconds and 3088K.
NOTE: SAS Institute Inc., SAS Campus Drive, Cary, NC USA 27513-2414

                        The SAS System        22:34 Sunday, March 19, 2000    1

    Analysis Variable : D

           Mean      Variance      Std Dev     Minimum      Maximum       Range
        -------------------------------------------------------------------------
        0.5002564    0.0320726    0.1790882   0.2900000    0.8700000    0.5800000
        -------------------------------------------------------------------------
                        The SAS System        22:34 Sunday, March 19, 2000    2

       Analysis Variable : N

               Mean        Sum     N    Minimum      Maximum       Range
           ----------------------------------------------------------------
           58.5000000  585.0000000  10   20.0000000  100.0000000   80.0000000
           ----------------------------------------------------------------
                        The SAS System        22:34 Sunday, March 19, 2000    3

OBS    SWMD    VAR_DLTA   SD_DLTA    VAR_E    PVA_SE   L95_CONF   U95_CONF   CHI_SQ

  1   0.50026      0          0     0.073056  227.784   0.50026    0.50026   4.39013
```

META-ANALYSIS RESULTS SUMMARY TABLE AND INTERPRETATION OF RESULTS FOR TRAINING EFFECTIVENESS EXAMPLE USED IN TABLES 2.1, 2.2, 2.4, AND 2.6

This section presents a summary table of the meta-analysis results for the training effectiveness example used in Tables 2.1, 2.2, 2.4, and 2.6. Text representing the interpretation of these results is also presented.

Interpretation of Results Presented in Table 2.7

Table 2.7 presents the results for the example used in Tables 2.1, 2.2, 2.4, and 2.6. The results of the meta-analysis show that the sample-weighted (corrected)

TABLE 2.7
Results of Overall Meta-Analysis for Training Effectiveness Example

Number of Data Points (k)	Total Sample Size	Sample-weighted Mean d	Corrected SD	% Variance due to Sampling Error	95% CI		χ^2
					L	U	
10	585	0.50	0.00	100.00	0.49	0.49	4.39

mean effect size was 0.50. One-hundred percent of the variance was accounted for by sampling error. This was consistent with both the absence of variability in the corrected mean d and the width of the confidence interval. The chi-square test for homogeneity also was not significant, further suggesting the absence of any moderators and supporting the conclusion that the data points are from the same population. In conclusion, the results indicate that the training program is moderately effective and its effectiveness is not influenced by any moderator variables or factors.

CONDUCTING MODERATOR ANALYSIS IN THE META-ANALYSES OF EFFECT SIZES

In general, variable Z is considered to be a moderator of the relationship between variables X and Y when the nature of this relationship is contingent upon the levels or values of Z. In the context of meta-analysis, a moderator variable is defined as any variable that, by its inclusion in the analysis accounts for, or helps explain, more variance than would otherwise be the case. This is somewhat different from the use of the term *moderator* in multiple regression. In multiple regression, a moderator is a variable that, although having a negligible correlation with a criterion, interacts with another variable to enhance the predictability of the criterion variable (Cohen & Cohen, 1983). In the meta-analysis of effect sizes, when sufficient variance remains in the population (i.e., corrected δ) after the specified corrections have been made, the presence of one or more moderator variables is suspected. Alternately, various moderator variables may be suggested by theory. Thus, the decision to search or test for moderators may be driven either theoretically or empirically. So, for instance, in our training effectiveness example, we may have some theoretical or conceptual basis to hypothesize that the effectiveness of the training program may be a function of how the dependent variable is operationalized—either using objective or subjective measures. And on this basis, we will then make a decision to test for criterion type as a moderator.

Alternately, this decision could be made primarily on the basis of the amount of variance accounted for and the degree of variability in δ after all the specified corrections have been made. Thus, in meta-analysis, a moderator variable is identified if: (a) the effect size variance is lower in the subsets or sublevels of the variable than the aggregate level, and/or (b) the average effect size varies from subset to subset. In short, in meta-analysis, if large differences are found between sublevels of a given variable, then that variable can be considered to be a moderator variable.

A number of statistical tools or methods have been used to make decisions about whether any of the observed variance in d or δ values across studies is real and not solely the result of artifacts. Specifically, these methods are used to search and test for the presence and effect of potential or hypothesized moderator variables. Examples of these methods include (a) using the 75% rule, (b) testing for significant differences between the corrected mean effect sizes of sublevels of the hypothesized moderator variable, and (c) using a chi-square test of homogeneity to determine whether the unexplained residual variance is significantly different from zero, or stated differently, to test whether the observed effect sizes are from the same population or different subpopulations (i.e., different levels of the moderator).

The 75% rule (which is more commonly associated with and used in the meta-analysis of correlations), states that if 75% or more of the variance can be accounted for by the specified corrections, then it can be concluded that all of the variance is due to artifacts since the remaining 25% is likely to be because of uncorrected artifacts such as computational and typographical errors. On the other hand, if the variation in effect sizes is determined not to be artifactual but "real", then moderator variables may be operating. There are generally two types of moderators: those that are theoretically predicted, and those that are unsuspected. For those moderators that have been theoretically postulated, one could use statistical significance tests comparing the mean effect sizes of the various levels of the moderator variable (Alliger, 1995; see also Aguinis & Pierce 1998). For unsus-

pected or unpredicted moderating variables, one must examine the residual variation left over after the specified corrections have been made to determine whether
it is different from zero. Consequently, a chi-square test can be used to try to detect these moderators by testing for the homogeneity of the true effect sizes across
studies.

Because these three methods answer somewhat different questions and also
use different decision rules, the results obtained from them may or may not be
congruent (e.g., see Arthur, Barrett, & Alexander, 1991). Consequently, it is usually advisable to use at least two or more of these methods in the search and test
for moderators.

On a cautionary note, the use of statistical significance tests to test for moderators in meta-analysis may be misguided (see Hunter & Schmidt, 1990a; Schmidt,
1992). It has been argued that the use of significance tests in primary research
under conditions of low power (which is almost universal in research literatures)
is the very reason why research literatures appear contradictory and confusing in
the first place (see Cortina & Dunlap, 1997; Hagen, 1997; Schmidt, 1992, 1996;
Schmidt & Hunter, 1978; Thompson, 1996; Wilkinson et al., 1999, for a discussion of this issue)—and consequently, the reason why meta-analysis is necessary
to make sense of these literatures. It is, therefore, illogical to introduce this same
problem to meta-analysis itself. For instance, Hunter and Schmidt (1990a) do not
endorse the chi-square test for homogeneity because "this significance test . . .
asks the wrong question. Significant variation may be trivial in magnitude, and
even nontrivial variation may still be due to research artifacts" (p. 110). They also
go on to caution that "if the meta-analysis has very many studies [data points], it
has very high statistical power and will therefore reject the null hypothesis, given
even a trivial amount of variation across studies" (p. 112).

Analysis of Moderator Variables

The analysis of moderator variables using the SAS PROC MEANS procedure is
relatively simple. The level of the moderator variable is first coded and entered
along with the d and sample size for each study (e.g., objective and subjective
criteria in Table 2.1). An *IF* statement can then be used to select only objective
criterion studies (*IF CRITERIA='OBJ';*) or subjective criterion studies (*IF*

TABLE 2.8
SAS Statements for Conducting a Moderator Analysis for Hypothetical Training Effectiveness Data With Criterion Type (Objective Criteria) as a Moderator Variable for the Meta-Analysis of Effect Sizes

```
001   DATA OBSERVED;
002   INPUT STUDY D N CRITERIA $;
003   CARDS;
004   1    .34   100   OBJ
005   2    .30   40    SBJ
006   3    .41   65    SBJ
007   4    .70   75    SBJ
008   5    .54   85    OBJ
009   6    .87   50    OBJ
010   7    .32   20    OBJ
011   8    .77   25    SBJ
012   9    .29   35    SBJ
013   10   .47   90    OBJ
014   ;
015   DATA SMP_WGT1;
016     SET OBSERVED;            ***********************************************************;
017   IF CRITERIA='OBJ';     **      FOR SUBJECTIVE CRITERIA, CHANGE 'OBJ' TO 'SBJ'     **;
018                          **      HERE AND IN DATA SMP_WGT2 AND RERUN               **;
019                          ***********************************************************;
020   PROC MEANS MEAN VAR STD MIN MAX RANGE VARDEF=WEIGHT;
021       VAR D;
022       WEIGHT N;
023       OUTPUT OUT=ADATA MEAN=SWMD VAR=VAR_D;
024
025   DATA SMP_WGT2;
026     SET OBSERVED;            ***********************************************************;
027   IF CRITERIA='OBJ';     **      FOR SUBJECTIVE CRITERIA, CHANGE 'OBJ' TO 'SBJ'     **;
028                          ***********************************************************;
029   PROC MEANS MEAN SUM N MIN MAX RANGE;
030       VAR N;
031       OUTPUT OUT=BDATA MEAN=MEAN_N SUM=N N=K;
032
033   DATA FINALE;
034     MERGE ADATA BDATA;
035       VAR_E = ((MEAN_N - 1) / (MEAN_N - 3)) * (4 / MEAN_N) * (1 + ((SWMD ** 2) / 8));
036       VAR_DLTA = VAR_D - VAR_E;
037     IF VAR_DLTA LT 0 THEN VAR_DLTA = 0;
038       PVA_SE = 100 * (VAR_E / VAR_D);
039       SD_DLTA = SQRT (VAR_DLTA);
040       L95_CONF = SWMD - (1.96 * SD_DLTA);
041       U95_CONF = SWMD + (1.96 * SD_DLTA);
042       CHI_SQ = K * VAR_D / VAR_E;
043   PROC PRINT;
044       VAR SWMD VAR_DLTA SD_DLTA VAR_E PVA_SE
045                       L95_CONF U95_CONF CHI_SQ ;
```

Note. The SAS code above is for objective criteria only. To run the meta-analysis for subjective criteria, the level of the moderator variable will have to be selected in the *IF* statements and the program rerun. (The SAS code for all the moderators used in this example can be found at www.erlbaum.com.) An alternative is to run the analyses for both levels of the moderator in the same program by repeating or cutting and pasting the subjective criteria code to the end of the objective criteria code. Running all levels of the moderator in the same program can also be accomplished by using *PROC SORT* or *WHERE* statements.

TABLE 2.9
SAS Printout for the Meta-Analysis of Effect Sizes—Moderator Analysis for Hypothetical Training Effectiveness Data With Criterion Type (Objective Criteria) as a Moderator Variable

```
1                          The SAS System        23:10 Sunday, March 19, 2000
NOTE: Copyright (c) 1989-1996 by SAS Institute Inc., Cary, NC, USA.
NOTE: SAS (r) Proprietary Software Release 6.09  TS470

*******************************************************************
*                                                                 *
*         Welcome to the new SAS System, Release 6.09 Enhanced    *
*                                                                 *
*******************************************************************

NOTE: SAS system options specified are:
      SORT=4

NOTE: The initialization phase used 0.15 CPU seconds and 2041K.
1             OPTIONS LS=90;
2             DATA OBSERVED;
3               INPUT STUDY D  N  CRITERIA $;
4               CARDS;

NOTE: The data set WORK.OBSERVED has 10 observations and 4 variables.
NOTE: The DATA statement used 0.04 CPU seconds and 2679K.

15              ;
16            DATA SMP_WGT1;
17              SET OBSERVED;    *****************************************;
18              IF CRITERIA='OBJ';  ** FOR SUBJECTIVE CRITERIA, CHANGE 'OBJ' TO 'SBJ' *;
19                                  ** HERE AND IN DATA SMP_WGT1 AND RERUN        *;
20                                  *****************************************;

NOTE: The data set WORK.SMP_WGT1 has 5 observations and 4 variables.
NOTE: The DATA statement used 0.01 CPU seconds and 2683K.

21            PROC MEANS MEAN VAR STD MIN MAX RANGE VARDEF=WEIGHT;
22              VAR D;
23              WEIGHT N;
24              OUTPUT OUT=ADATA MEAN=SWMD  VAR=VAR_D;
25

NOTE: The data set WORK.ADATA has 1 observations and 4 variables.
NOTE: The PROCEDURE MEANS printed page 1.
NOTE: The PROCEDURE MEANS used 0.03 CPU seconds and 2888K.

26            DATA SMP_WGT2;
27              SET OBSERVED;    *****************************************;
28              IF CRITERIA='OBJ';  ** FOR SUBJECTIVE CRITERIA, CHANGE 'OBJ' TO 'SBJ' *;
29                                  *****************************************;
2                          The SAS System        23:10 Sunday, March 19, 2000

NOTE: The data set WORK.SMP_WGT2 has 5 observations and 4 variables.
NOTE: The DATA statement used 0.01 CPU seconds and 2888K.

30            PROC MEANS MEAN SUM N MIN MAX RANGE;
31              VAR N;
32              OUTPUT OUT=BDATA MEAN=MEAN_N  SUM=N  N=K;
33

NOTE: The data set WORK.BDATA has 1 observations and 5 variables.
NOTE: The PROCEDURE MEANS printed page 2.
NOTE: The PROCEDURE MEANS used 0.01 CPU seconds and 2888K.
```

(continued)

CRITERIA='SBJ';) into the analyses. The complete meta-analysis statistics are then recalculated for each level of the moderator variable. A moderator is identified by (a) a corrected variance that has a lower average in the subsets than

```
34            DATA FINALE;
35               MERGE  ADATA BDATA;
36                  VAR_E = ((MEAN_N - 1) / (MEAN_N - 3)) * (4 / MEAN_N) *
37                       (1 + ((SWMD ** 2) / 8));
38                  VAR_DLTA = VAR_D - VAR_E;
39               IF VAR_DLTA LT 0 THEN VAR_DLTA = 0;
40                  PVA_SE = 100 * (VAR_E / VAR_D);
41                  SD_DLTA = SQRT (VAR_DLTA);
42                  L95_CONF = SWMD - (1.96 * SD_DLTA);
43                  U95_CONF = SWMD + (1.96 * SD_DLTA);
44                  CHI_SQ = K * VAR_D / VAR_E;
NOTE: The data set WORK.FINALE has 1 observations and 17 variables.
NOTE: The DATA statement used 0.04 CPU seconds and 3005K.

45            PROC PRINT;
46               VAR  SWMD  VAR_DLTA  SD_DLTA  VAR_E  PVA_SE
47                  L95_CONF  U95_CONF  CHI_SQ;
NOTE: The PROCEDURE PRINT printed page 3.
NOTE: The PROCEDURE PRINT used 0.01 CPU seconds and 3088K.

NOTE: The SAS session used 0.33 CPU seconds and 3088K.
NOTE: SAS Institute Inc., SAS Campus Drive, Cary, NC USA 27513-2414

                              The SAS System       23:10 Sunday, March 19, 2000   1

     Analysis Variable : D

           Mean        Variance      Std Dev      Minimum      Maximum      Range
     ------------------------------------------------------------------------------
         0.4988406     0.0297668    0.1725305    0.3200000    0.8700000    0.5500000
     ------------------------------------------------------------------------------
                              The SAS System       23:10 Sunday, March 19, 2000   2

     Analysis Variable : N

           Mean          Sum N      Minimum      Maximum      Range
     ----------------------------------------------------------------
         69.0000000   345.0000000    5   20.0000000  100.0000000   80.0000000
     ----------------------------------------------------------------
                              The SAS System       23:10 Sunday, March 19, 2000   3

  OBS    SWMD    VAR_DLTA   SD_DLTA    VAR_E     PVA_SE   L95_CONF   U95_CONF   CHI_SQ

   1    0.49884      0          0     0.061586   206.894   0.49884    0.49884   2.41670
```

Note. This printout is for objective criteria only. The program will have to be rerun for subjective criteria by changing the level of the moderator variable in the *IF* statements. (The SAS code for all the moderators used in this example can be found at www.erlbaum.com.) An alternative is to run the analyses for both levels of the moderator in the same program by repeating or cutting and pasting the subjective criteria code to the end of the objective criteria code. Running all levels of the moderator in the same program can also be accomplished by using *PROC SORT* or *WHERE* statements.

for the data as a whole, and (b) an average d (or δ, if we make corrections) that varies from subset to subset.

Alternatively, one could use *PROC SORT* coupled with *BY* statements, or even *WHERE* statements to run these analyses. The advantage to *PROC SORT*s is that one does not have to separately rerun the analysis for each level of the moderator because analyses of all the different levels can be accomplished within a

Table 2.10
Meta-Analysis Results of Criterion Type Moderator Analysis
for Training Effectiveness Example

VARIABLE	No. of Data Points (k)	Total Sample Size	Sample-weighted Mean d	Corrected SD	% Variance due to Sampling Error	95% CI		χ^2
						L	U	
Overall	10	585	0.50	0.00	100.00	0.50	0.50	4.39
Criterion-Type								
Objective Criteria	5	345	0.50	0.00	100.00	0.50	0.50	2.42
Subjective Criteria	5	240	0.50	0.00	100.00	0.50	0.50	1.97

Note. The chi-square is not significant. At $p = .05$, the critical chi-square value at k-1 degrees of freedom is 16.02 for $df = 9$, and 9.49 for $df = 4$.

single run. The disadvantage is that although it is not that difficult to master, one has to be familiar with *PROC SORT*s in general, and the use of multiple *PROC SORT* statements within the same program in particular. For the sake of simplicity and clarity, we have chosen to limit our presentation and examples to the use of *IF* statements.

SAS code (using *IF* statements) necessary to run a moderator analysis (criterion type—objective vs. subjective) on the hypothetical training effectiveness data is presented in Table 2.8. The SAS printout is presented in Table 2.9, and a summary of the meta-analysis results is presented in Table 2.10. In general, a moderator is identified by (a) a corrected variance that has a lower average in the subsets than for the data as a whole, and (b) an average d (or δ, if we make corrections) that varies from subset to subset. Consistent with the results obtained for the overall analysis and its interpretation (see Table 2.7), these results of the moderator analysis show that criterion type does not operate as a moderator.

Fully Hierarchical Moderator Analysis

Fully hierarchical moderator analysis is basically an investigation of interactions among the moderators. Thus, we might be interested in investigating whether training evaluation studies that use speeded tasks that are tested using recognition tests result in different effectiveness outcomes than those that use speeded tasks that are tested using recall tests. Fully hierarchical moderator analyses are technically superior to partial or incomplete moderator analyses. However, the most frequent, albeit not the best practice in testing for moderators, is to first include all the studies in an overall meta-analysis and then break the studies down by one moderator (e.g., test type; see Table 2.13), and then by another moderator (e.g., task type), and so on. This approach to moderator analysis is not fully hierarchical because the moderators are not considered in combination, but instead, they are considered independently of one another. Failure to consider moderator variables in combination can result in major errors of interpretation since one fails to take into account interaction and confounding effects (Hunter & Schmidt, 1990a).

The implementation of a fully hierarchical moderator analysis first calls for the creation of a matrix that represents all levels of all moderators. The data points are then distributed across the cells of the matrix. Thus, the feasibility of fully hierarchical moderator analyses is primarily a function of the number of data points in each cell and is constrained by the total number of data points, the number of moderators, and the number of levels of the moderators. Although technically superior, fully hierarchical moderator analyses are not very common in the meta-analysis literature because typically, when the meta-analytic data are disaggregated to this level, the number of data points in each cell is usually quite small, and the meta-analysis of such small numbers of data points is very questionable since it raises major and serious concerns about the stability and interpretability of meta-analytic estimates. For instance, in the example presented in Table 2.10, disaggregating the data to just two levels of a single moderator resulted in five data points per cell. And in meta-analytic terms, five (and even ten) data points is a relatively small sample of data points.

Specifically, although this goal is rarely met, a meta-analysis ideally calls for several hundred data points. This issue is further exacerbated by moderator and even more so by fully hierarchical moderator analysis because disaggregating variables down to sublevels reduces, sometimes drastically, the number of effect sizes to be meta-analyzed to a relatively small number. As the number of studies in a meta-analysis decreases, the likelihood of sampling error increases. Consequently, when one corrects for sampling error, it becomes more likely to obtain favorable than unfavorable results. Meta-analyses of small numbers of

TABLE 2.11
SAS Statements for Conducting a Fully Hierarchical Moderator Analysis
for Hypothetical Training Effectiveness Data With Test Type (Recognition)
and Task Type (Speed) as Moderator Variables
for the Meta-Analysis of Effect Sizes

```
001   DATA OBSERVED;
002   INPUT STUDY D N TASK $ TEST $;
003   CARDS;
004   1    0.20  15   ACCURACY   RECALL
005   2    0.32  140  ACCURACY   RECALL
006   3    0.88  145  SPEED      RECOGN
007   4    0.95  35   SPEED      RECOGN
008   5    0.51  45   SPEED      RECOGN
009   6    0.43  135  ACCURACY   RECOGN
010   7    0.08  180  ACCURACY   RECALL
011   8    0.01  50   SPEED      RECALL
012   9    0.17  65   ACCURACY   RECOGN
013   10   0.50  70   SPEED      RECALL
014   11   0.14  80   ACCURACY   RECALL
015   12   0.05  135  ACCURACY   RECOGN
016   13   0.16  15   SPEED      RECOGN
017   14   0.14  120  SPEED      RECALL
018   15   0.61  60   ACCURACY   RECOGN
019   16   0.08  130  SPEED      RECALL
020   17   0.99  230  SPEED      RECOGN
021   18   0.21  90   SPEED      RECALL
022   19   0.16  180  SPEED      RECALL
023   20   0.74  225  ACCURACY   RECOGN
024   21   0.80  98   SPEED      RECOGN
025   22   0.42  80   SPEED      RECOGN
026   23   0.84  15   SPEED      RECOGN
027   24   0.07  215  ACCURACY   RECALL
028   25   0.37  35   SPEED      RECALL
029   26   0.99  110  SPEED      RECOGN
030   27   0.53  110  SPEED      RECALL
```

(continued)

studies are biased in favor of obtaining "positive" results. Thus, they are to be discouraged, and if unavoidable, should be cautiously interpreted because they could be very misleading. Although there is no magic cutoff as to the minimum number of studies, it is conceptually difficult to make a case that a meta-analysis of five, or even ten data points represents a *population* parameter.

The SAS code required to run a fully hierarchical moderator analysis is quite simple. Specifically, we can run a fully hierarchical moderator analysis by simultaneously analyzing additional moderator variables by using joint or compound IF statements (e.g., *IF TEST='RECALL' AND TASK='SPEED';*). SAS code necessary to run a fully hierarchical moderator analysis on another hypothetical training effectiveness data set is presented in Table 2.11. The two

031	28	0.56	40	ACCURACY	RECOGN
032	29	0.51	240	SPEED	RECOGN
033	30	0.04	115	SPEED	RECOGN
034	31	0.34	50	SPEED	RECOGN
035	32	0.26	320	ACCURACY	RECALL
036	33	0.62	45	SPEED	RECALL
037	34	0.82	10	SPEED	RECOGN
038	35	0.42	10	SPEED	RECOGN
039	36	0.56	35	SPEED	RECOGN
040	37	0.86	25	SPEED	RECOGN
041	38	0.13	20	SPEED	RECALL
042	39	0.43	40	SPEED	RECOGN
043	40	0.50	45	ACCURACY	RECOGN
044	41	0.33	220	ACCURACY	RECOGN
045	42	0.24	20	ACCURACY	RECOGN
046	43	0.22	45	SPEED	RECALL
047	44	0.14	20	ACCURACY	RECALL
048	45	0.27	20	SPEED	RECALL
049	46	0.32	80	ACCURACY	RECOGN
050	47	0.38	100	SPEED	RECOGN
051	48	0.33	120	SPEED	RECALL
052	49	0.16	20	ACCURACY	RECALL
053	50	0.69	15	SPEED	RECALL
054	51	0.89	15	SPEED	RECOGN
055	52	0.85	60	SPEED	RECOGN
056	53	0.46	55	SPEED	RECALL
057	54	0.40	20	SPEED	RECALL
058	55	0.34	15	SPEED	RECALL
059	56	0.05	120	ACCURACY	RECALL
060	57	0.95	15	SPEED	RECOGN
061	58	0.66	75	SPEED	RECOGN
062	59	0.09	30	ACCURACY	RECALL
063	60	0.18	60	ACCURACY	RECALL
064	61	0.05	110	SPEED	RECALL
065	62	0.20	155	SPEED	RECOGN
066	63	0.57	90	SPEED	RECOGN
067	64	0.10	95	ACCURACY	RECALL
068	65	0.27	105	SPEED	RECOGN
069	66	0.15	20	SPEED	RECOGN
070	67	0.02	150	ACCURACY	RECALL
071	68	0.09	80	ACCURACY	RECALL
072	69	0.09	100	ACCURACY	RECOGN
073	70	0.16	45	SPEED	RECALL
074	71	0.53	80	ACCURACY	RECOGN
075	72	0.38	75	SPEED	RECALL
076	73	0.29	25	SPEED	RECALL
077	74	0.51	50	ACCURACY	RECOGN
078	75	0.34	105	SPEED	RECOGN
079	76	0.06	25	ACCURACY	RECALL
080	77	0.04	20	ACCURACY	RECOGN
081	78	0.24	40	ACCURACY	RECALL
082	79	0.70	70	SPEED	RECOGN
083	80	0.10	25	ACCURACY	RECOGN
084	81	0.56	20	SPEED	RECOGN
085	82	0.66	60	SPEED	RECOGN
086	83	0.81	40	SPEED	RECOGN
087	84	0.16	90	ACCURACY	RECALL
088	85	0.48	80	SPEED	RECOGN
089	86	0.35	115	ACCURACY	RECOGN
090	87	0.00	110˙	SPEED	RECALL

(continued)

```
091   88   0.11   190   SPEED      RECOGN
092   89   0.07   220   ACCURACY   RECALL
093   90   0.59   35    SPEED      RECALL
094   91   0.97   95    SPEED      RECOGN
095   92   0.91   20    SPEED      RECOGN
096   93   0.12   25    ACCURACY   RECALL
097   94   0.03   95    ACCURACY   RECALL
098   95   0.97   40    ACCURACY   RECOGN
099   96   0.69   80    SPEED      RECOGN
100   97   0.27   15    SPEED      RECALL
101   98   0.10   260   ACCURACY   RECALL
102   99   0.20   135   SPEED      RECALL
103   100  0.67   15    SPEED      RECOGN
104   ;
105   DATA SMP_WGT1;
106     SET OBSERVED;
107   IF TEST='RECOGN' AND TASK='SPEED';
108
109   PROC MEANS MEAN VAR STD MIN MAX RANGE VARDEF=WEIGHT;
110         VAR D;
111         WEIGHT N;
112         OUTPUT OUT=ADATA MEAN=SWMD VAR=VAR_D;
113
114   DATA SMP_WGT2;
115     SET OBSERVED;
116   IF TEST='RECOGN' AND TASK='SPEED';
117
118   PROC MEANS MEAN SUM N MIN MAX RANGE;
119         VAR N;
120         OUTPUT OUT=BDATA MEAN=MEAN_N SUM=N N=K;
121
122   DATA FINALE;
123     MERGE ADATA BDATA;
124       VAR_E = ((MEAN_N - 1) / (MEAN_N - 3)) * (4 / MEAN_N) * (1 + ((SWMD ** 2) / 8));
125       VAR_DLTA = VAR_D - VAR_E;
126     IF VAR_DLTA LT 0 THEN VAR_DLTA = 0;
127       PVA_SE = 100 * (VAR_E / VAR_D);
128       SD_DLTA = SQRT (VAR_DLTA);
129       L95_CONF = SWMD - (1.96 * SD_DLTA);
130       U95_CONF = SWMD + (1.96 * SD_DLTA);
131       CHI_SQ = K * VAR_D / VAR_E;
132   PROC PRINT;
133         VAR SWMD VAR_DLTA SD_DLTA VAR_E PVA_SE
134               L95_CONF U95_CONF CHI_SQ;
```

Note. The above SAS code is for recognition tests and speed tasks only. To run the meta-analysis for the other moderators, the joint levels of the moderator variables will have to be selected in the *IF* statements and the program rerun. (The SAS code for all the moderators can be found at www.erlbaum.com.) An alternative is to run the analyses for all combinations of the moderator variables in the same program by repeating or cutting and pasting the code for the other moderator variable levels to the end of the code presented in Table 2.11. Running all levels of the moderator in the same program can also be accomplished by using *PROC SORT* or *WHERE* statements.

TABLE 2.12

SAS Printout for the Meta-Analysis of Effect Sizes—Fully Hierarchical Moderator Analysis for Hypothetical Training Effectiveness Data With Test Type (Recognition) and Task Type (Speed) as the Moderator Variables

```
1                               The SAS System        00:43 Tuesday, March 21, 2000

NOTE: Copyright (c) 1989-1996 by SAS Institute Inc., Cary, NC, USA.
NOTE: SAS (r) Proprietary Software Release 6.09  TS470

***********************************************************************
*                                                                     *
*          Welcome to the new SAS System, Release 6.09 Enhanced       *
*                                                                     *
***********************************************************************

NOTE: SAS system options specified are:
      SORT=4
NOTE: The initialization phase used 0.17 CPU seconds and 2041K.
1          OPTIONS LS=90;
2          DATA OBSERVED;
3            INPUT STUDY D  N  TASK $  TEST $;
4            CARDS;

NOTE: The data set WORK.OBSERVED has 100 observations and 5 variables.
NOTE: The DATA statement used 0.06 CPU seconds and 2679K.

105        ;
106        DATA SMP_WGT1;
107          SET OBSERVED;
108        IF TEST='RECOGN' AND TASK='SPEED';
109
110

NOTE: The data set WORK.SMP_WGT1 has 37 observations and 5 variables.
NOTE: The DATA statement used 0.01 CPU seconds and 2683K.

111        PROC MEANS MEAN VAR STD MIN MAX RANGE VARDEF=WEIGHT;
112          VAR D;
113          WEIGHT N;
114          OUTPUT OUT=ADATA MEAN=SWMD  VAR=VAR_D;
115

NOTE: The data set WORK.ADATA has 1 observations and 4 variables.
NOTE: The PROCEDURE MEANS printed page 1.
NOTE: The PROCEDURE MEANS used 0.03 CPU seconds and 2888K.

116        DATA SMP_WGT2;
117          SET OBSERVED;
118        IF TEST='RECOGN' AND TASK='SPEED';
119

2                               The SAS System        00:43 Tuesday, March 21, 2000

NOTE: The data set WORK.SMP_WGT2 has 37 observations and 5 variables.
NOTE: The DATA statement used 0.01 CPU seconds and 2888K.

120        PROC MEANS MEAN SUM N MIN MAX RANGE;
121          VAR N;
122          OUTPUT OUT=BDATA MEAN=MEAN_N  SUM=N  N=K;
123

NOTE: The data set WORK.BDATA has 1 observations and 5 variables.
NOTE: The PROCEDURE MEANS printed page 2.
NOTE: The PROCEDURE MEANS used 0.01 CPU seconds and 2888K.
```

(continued)

```
124        DATA FINALE;
125          MERGE  ADATA BDATA;
126            VAR_E = ((MEAN_N - 1) / (MEAN_N - 3)) * (4 / MEAN_N) *
127                   (1 + ((SWMD ** 2) / 8));
128          VAR_DLTA = VAR_D - VAR_E;
129        IF VAR_DLTA LT 0 THEN VAR_DLTA = 0;
130          PVA_SE = 100 * (VAR_E / VAR_D);
131          SD_DLTA = SQRT (VAR_DLTA);
132          L95_CONF = SWMD - (1.96 * SD_DLTA);
133          U95_CONF = SWMD + (1.96 * SD_DLTA);
134          CHI_SQ = K * VAR_D / VAR_E;
NOTE: The data set WORK.FINALE has 1 observations and 17 variables.
NOTE: The DATA statement used 0.04 CPU seconds and 3005K.

135        PROC PRINT;
136          VAR  SWMD VAR_DLTA SD_DLTA VAR_E  PVA_SE
137               L95_CONF  U95_CONF  CHI_SQ;
NOTE: The PROCEDURE PRINT printed page 3.
NOTE: The PROCEDURE PRINT used 0.02 CPU seconds and 3088K.

NOTE: The SAS session used 0.37 CPU seconds and 3088K.
NOTE: SAS Institute Inc., SAS Campus Drive, Cary, NC USA 27513-2414
```

```
                               The SAS System      00:43 Tuesday, March 21, 2000   1

    Analysis Variable : D

          Mean        Variance       Std Dev      Minimum       Maximum        Range
      -----------------------------------------------------------------------------------
      0.5704025     0.0905813      0.3009673     0.0400000     0.9900000     0.9500000
      -----------------------------------------------------------------------------------
                               The SAS System      00:43 Tuesday, March 21, 2000   2

    Analysis Variable : N

          Mean         Sum    N      Minimum       Maximum        Range
      --------------------------------------------------------------------------
      73.1891892     2708.00  37   10.0000000    240.0000000   230.0000000
      --------------------------------------------------------------------------
                               The SAS System      00:43 Tuesday, March 21, 2000   3

  OBS     SWMD     VAR_DLTA    SD_DLTA    VAR_E      PVA_SE    L95_CONF   U95_CONF    CHI_SQ

   1     0.57040   0.032085    0.17912   0.058496   64.5787    0.21932    0.92148   57.2944
```

Note. This printout is for recognition tests and speed tasks only. The program will have to be rerun for the other combinations of the moderator variable (see Table 2.13) by changing the levels of the joint moderator variables in the *IF* statements. An alternative is to run the analyses for all combinations of the moderator in the same program by repeating or cutting and pasting the code for the other moderator variables to the end of the code in Tables 2.11 and 2.12. Running all combinations of the moderator in the same program can also be accomplished by using *PROC SORT* or *WHERE* statements.

moderators are the nature of the learning test (recognition or recall) and task type (speed or accuracy). The SAS printout for Table 2.11 is presented in Table 2.12, and a summary of the meta-analysis results is presented in Table 2.13.

TABLE 2.13

Meta-Analysis Results for Fully Hierarchical Moderator Analysis for Training Effectiveness Example

VARIABLE	No. of Data Points (k)	Total Sample Size	Sample-weighted Mean d	Corrected SD	% Variance due to Sampling Error	95% CI		χ^2
						L	U	
Overall	100	8133	0.35	0.18	62.57	0.01	0.69	159.83*
Test Type								
Recognition	54	4163	0.51	0.17	64.74	0.17	0.84	83.41*
Recall	46	3970	0.18	0.00	100.00	0.18	0.18	20.64
Task Type								
Speed	62	4398	0.45	0.18	64.07	0.09	0.80	96.78*
Errors	38	3735	0.23	0.05	93.57	0.12	0.34	40.61
Test/Task Type								
Recognition/Speed	37	2708	0.57	0.18	64.58	0.22	0.91	57.29*
Recognition/Errors	17	1455	0.40	0.09	86.65	0.23	0.58	19.62
Recall/Speed	25	1690	0.25	0.00	100.00	0.24	0.24	12.92
Recall/Errors	21	2280	0.13	0.00	100.00	0.13	0.13	4.27

Note. $*p < .05$, $df = k\text{-}1$.

Interpretation of Results Presented in Table 2.13

Results of the meta-analysis presented in Table 2.13 show that the sample-weighted mean effect size was 0.35 with a standard deviation of 0.18. Approximately 63% of the variance was accounted for by sampling error. The 95% confidence interval about the mean effect size was fairly large (0.01 to 0.69) and the chi-square test was significant, suggesting the operation of possible moderators. Therefore, although the training program is generally moderately effective, its effectiveness appears to be influenced by moderators.

This was supported by the results of a subsequent fully hierarchical moderator analysis of test and task type. These results indicated that for test type, trainee performance was higher when performance was assessed using a recognition ($\overline{d} = 0.51$) instead of a recall test ($\overline{d} = 0.18$). Likewise, for task type, performance was higher on speeded tests ($\overline{d} = 0.45$) in contrast to tests of accuracy ($\overline{d} = 0.23$). A further breakdown of the data by both test and task type indicated that performance was highest when speeded tasks were used in conjunction with recognition tests ($\bar{d} = 0.57$). The least effective performance was obtained for accuracy-based recall tests ($\bar{d} = 0.13$). It is also noteworthy that the results of the fully hierarchical moderator analyses further indicate that, with the exception of maybe recognition/speed tasks, there do not appear to be any additional moderators operating; substantial amounts of sampling error variance (greater than 85%) were accounted for, the chi-square tests of homogeneity were not significant, and the 95% confidence intervals were relatively small and did not include zero.

In conclusion, these results indicate that the training program is, in general, moderately effective and furthermore, that its effectiveness is influenced by both the test and task type used to operationalize trainee performance. Specifically, the highest levels of trainee performance are obtained with speed-based recognition tests, followed by accuracy-based recognition, speed-based recall, and accuracy-based recall tests which resulted in fairly small levels of trainee performance.

SUMMARY

This chapter presented the SAS PROC MEANS procedure in the meta-analysis of effect sizes, focusing primarily on the Glassian approach. With this approach, the two summary statistics of interest are the sample-weighted mean d and its variance. The Glassian approach does not correct for any statistical artifacts. However, Hunter and Schmidt (1990a) have presented extensions of the Glassian approach which correct for sampling error, and even criterion reliability, in the meta-analysis of ds. We presented these extensions and the SAS code required to implement them.

Specifically, we focused on (a) the estimation of sampling error variance, which cumulated in the computation of the percent variance accounted for by sampling error, and (b) the computation of confidence intervals, which are used to assess the accuracy of the mean effect size. We also discussed (c) correcting the mean d for statistical artifacts, namely criterion reliability, and (d) a chi-square test for homogeneity, which can be used to assess and test for the effects of moderator variables.

In relation to the chi-square test for homogeneity, we also (e) reviewed and discussed the issue of conducting moderator analysis in the meta-analysis of effect sizes. The logical inconsistency of applying statistical significance tests in the analysis of moderators in meta-analysis was discussed. We also noted and acknowledged that despite this, it is not an uncommon practice. In discussing moderator analysis, we also (f) noted the distinction between fully hierarchical and partial or incomplete moderator analysis, and highlighted the advantages of the former—fully hierarchical moderator analysis—over the latter. Although technically superior, fully hierarchical analyses are performed very infrequently in the extant literature. This is because the implementation of a fully hierarchical moderator analysis first calls for the creation of a matrix that represents all levels of all the moderators. Thus, the feasibility of these analyses is constrained by the number of data points in each cell.

Finally, the above issues were discussed within the context of specific examples which were illustrated with tables presenting the actual SAS code along with representations of the actual SAS printouts. Coupled with these, we presented tables summarizing the meta-analysis results of our examples, along with interpretations of these results. The SAS code and data used in the examples presented in this chapter can be downloaded at www.erlbaum.com. Chapter 3 presents the use of the SAS PROC MEANS procedure to run a meta-analysis of correlations (rs) with a focus on the Hunter and Schmidt (1990a) approach.

Meta-Analysis
Of Correlations

INTRODUCTION AND OVERVIEW

This chapter presents SAS code that can be used to conduct a meta-analysis of correlations with an emphasis on the Hunter and Schmidt approach, also referred to as the *validity generalization procedure* (Hunter & Schmidt, 1990a; Hunter et al., 1982; Schmidt & Hunter, 1977). The development of Hunter and Schmidt's validity generalization technique was in response to the *situational specificity hypothesis*.

In the 1940s and 1950s, it was widely believed that general cognitive ability predicted performance equally well for all persons in all situations. However, there was a shift later to a belief in the situational specificity hypothesis, which postulated that the validity of employment tests were not generalizable but instead were situationally specific. The resultant implication and practice was that employment tests needed to be validated each time they were used because jobs, and their embedded environments, varied from one situation to the next. The situational specificity hypothesis was predicated on the fact that the results of empirical validation studies showed the magnitude of validity coefficients to be highly variable across situations. Thus, the situational specificity hypothesis implied that validity generalization was impossible and empirical validation was required for each specific test use.

Validity generalization techniques were originally developed by Hunter and Schmidt to test the situational specificity hypothesis. The development of validity generalization procedures was independent of, but paralleled the concurrent development of similar procedures in clinical and counseling psychology by Rosenthal (Rosenthal, 1978, 1983, 1984; Rosenthal & Rubin, 1982, 1986) and

Glass (Glass, 1976; Glass & Kliegl, 1983; Glass et al., 1981) to assess the effi-
cacy of clinical and counseling interventions and address the issue of spontaneous
recovery.

Hunter and Schmidt's validity generalization technique combines the results
of previous studies—using them as a sample—to make aggregate summary statements about the relationships and effects obtained from these studies and their
generalizability. The basic premise of validity generalization is that the evidence
for situational specificity is artifactual in nature. Specifically, the observed differences in test validities from one study or situation to the next are the result of certain statistical artifacts or errors and not real differences. Therefore, if the residual
variance is at or near zero after the variance due to these artifacts is identified and
removed, then the situational specificity hypothesis is rejected and validity generalization is accepted.

STATISTICAL ARTIFACTS

This chapter begins with a brief overview and discussion of the statistical artifacts
of interest in the meta-analysis of correlations and validity generalization. These
artifacts are as follows:

- Sampling error
- Error of measurement—independent (predictor) and dependent variable
  (criterion) unreliability
- Range variation and restriction in the independent and dependent vari-
  able
- Dichotomization of variables
- Imperfect construct validity in the independent variable
- Imperfect construct validity in the dependent variable
- Errors in data—computational and typographical errors

The Hunter and Schmidt procedure allows for the correction of some of these
artifacts, the most common of which are sampling error, measurement error, and
range restriction (Steiner, Lane, Dobbins, Schnur, & McConnell, 1991). So, in the
implementation of a meta-analysis of correlations (i.e., validity generalization),
studies are first collected and their results (rs) are extracted. For example, we
might have 100 correlations (validity coefficients) relating scores on tests of per-
ceptual speed to overall proficiency in clerical jobs. The average correlation along

with its variance across studies is then calculated. The variance is corrected by subtracting the amount due to sampling error. The mean correlation and variance are then corrected for additional specified statistical artifacts other than sampling error. This meta-analytic result is a fully corrected correlation with its associated corrected variance. If we account for 75% or more of the corrected variance, the situational specificity hypothesis is rejected and validity generalization is accepted. Alternately, if we account for 75% of the variance or less, or the corrected variance is still large, then the next step is to test for the influence of moderator variables. To do so, the correlations are categorized on basis of the levels of the hypothesized moderator variable and the meta-analysis repeated. If a variable acts as a moderator, the means of the subcategories should be different and their variation should be less than the variation for the overall aggregated data set. The next section briefly reviews commonly discussed statistical artifacts.

Sampling Error

Sampling error is considered the largest source of variation across studies. Primary study correlations vary randomly from the population value because of sampling error. The magnitude of sampling error is determined primarily by sample size. Small sample studies will manifest higher levels of sampling error than will larger sample studies.

Error of Measurement—Independent (Predictor) and Dependent Variable (Criterion) Unreliability

The second largest source of variation across studies in most research literatures or domains is variation in error of measurement across studies. Predictor and criterion unreliability will attenuate observed correlations or validities. Thus, to the extent that the measurement of the *predictor* (independent variable) and *criterion* (dependent variable) include measurement error, the study correlation will be lower than the true correlation. Furthermore, different studies often use different measures of the independent and dependent variables of interest. To the extent that these measures differ in their reliabilities, there will be differences in the observed correlations across these studies.

Range Variation and Restriction in the Independent and Dependent Variable

If the variance (or standard deviation) of the independent or dependent variable differs substantially from one study to the next, then there will be corresponding variability in the observed correlations from study to study. So if correlations from different studies are to be aggregated, then as with the other statistical arti-facts, differences in the correlations due to differences in variance or distribution of scores on the independent or dependent variable must be controlled. Range restriction—the extent to which independent or dependent variable scores in the study sample have a smaller variance than the population—will also systemati-cally reduce the observed study correlation.

Dichotomization of Variables

It is not uncommon for researchers to artificially dichotomize continuous vari-ables. Examples of this are pass–fail on a test or task, young–old in reference to age, and rich–poor in lieu of yearly income. Under these conditions, the point biserial correlation for the dichotomized variable will be attenuated compared to the correlation for the continuous variable. This situation is further compounded if both variables are artificially dichotomized resulting in a double reduction of the (biserial) correlation. Thus, the artificial dichotomization of continuous vari-ables results in a downward distortion in the mean correlation and an upward dis-tortion in the observed variation of correlations across studies (Hunter & Schmidt, 1990b). Furthermore, to the extent that dichotomization varies across studies, there will be an associated variability in the magnitude of the observed correla-tions.

Imperfect Construct Validity in the Independent Variable

If multiple studies that are to be compared differ in terms of the factor structures of tests used to operationalize the independent variable (e.g., global vs. facet mea-sures of job satisfaction; Scarpello & Campbell, 1983; Wanous & Reichers, 1996; Wanous, Reichers, & Hudy, 1997), the correlations obtained from these studies will vary accordingly. Thus, in the context of this artifact, it is conceivable that although multiple studies ostensibly allege to measure the same construct, differences in the operationalizations of the construct may result in the studies in actuality, investigating quite different constructs.

Imperfect Construct Validity in the Dependent Variable

The case for imperfect construct validity in the dependent variable is the same as that for the independent variable. Specifically, if multiple studies that are to be compared differ in regard to the factor structures of measures used to operationalize the dependent variable, then the correlations obtained from these studies will vary accordingly. Differences in criterion factor structure may result from criterion contamination or deficiency.

Errors in Data—Computational and Typographical Errors in Data

Study validities can also vary as a result of errors in the data. Thus, reported study validities can differ from actual study validities as a result of a wide range of errors including, but not limited to computational and typographical errors, inaccuracies in coding data, and transcription and printing errors. Although they do not present any empirical data in its support, Hunter and Schmidt (1990a, p. 45) noted that these errors can be very large in magnitude. Sometimes a few of these errors are caught and corrected with subsequent errata. However, our suspicion is that these represent a very small percentage.

Hunter and Schmidt's approach to the meta-analysis of correlations (validity generalization) typically corrects for the first three statistical artifacts (sampling error, measurement error, and range restriction), and sometimes the fourth (dichotomization of variable). The remaining artifacts are not corrected for because they are difficult, if not impossible, to quantify. This serves as the basis for the 75% rule. Specifically, it is posited that if three or fewer artifacts account for 75% of the variance in the observed correlations, then other artifacts will most likely account for the remaining variance.

It is interesting to note that in their review of meta-analyses in organizational behavior and human resource management, Steiner et al. (1991) reported that in studies using the Hunter and Schmidt approach, 100% corrected for sampling error, 69.2% corrected for unreliability in the dependent variable, 50.0% corrected for unreliability in the independent variable, and 11.5% corrected for range re-striction in the independent variable.

In summary, this chapter presents SAS programming statements necessary to conduct a full Hunter and Schmidt meta-analysis with correction for multiple arti-facts. The procedures presented here conform to those outlined in Hunter and Schmidt (1990a) for correlational data. Variations of the Hunter and Schmidt meta-analysis (e.g., Hedges, 1989; Raju & Burke, 1983; Raju, Burke, Normand, & Langlois, 1991) are not covered or presented in this volume although the SAS statements can easily be adapted to them.

INDIVIDUALLY CORRECTED VERSUS ARTIFACT DISTRIBUTION-BASED CORRECTIONS

There are several ways in the Schmidt and Hunter approach in which study coef-ficients can be corrected for artifactual variance. First, each study can be *individu-ally* corrected for attenuating artifacts such as measurement error (predictor and criterion unreliability). However, this requires that attenuating artifact data be available for each single study included in the meta-analysis. Because this infor-mation is usually reported sporadically in the primary literature, a second, more common approach is to base corrections on *distributional* data. This latter ap-proach is accomplished by compiling the measurement attenuation data from the primary studies that report this data, and using it to form artifact distributions. These distributions are subsequently used to make specified meta-analytic correc-tions.

The use of the "mixed case" in artifact distribution corrections would occur in situations where every study presents information about at least one common artifact, however, information on other artifacts is missing from many (but not all) of the studies. For example, all studies present information on predictor reli-ability but only some studies present information on criterion reliability. In this situation, the data points are first individually corrected for the artifacts for which there is complete information (in our example, predictor reliability). Next, the partially corrected data points are meta-analyzed. Finally, the results of the in-terim meta-analysis are corrected using artifact distributions (in this example, criterion reliability) resulting in the meta-analytic result of a fully corrected correlation.

Because distributional procedures are more common than individually corrected or mixed case procedures, this chapter focuses on only full distributional procedures. The interested reader can refer to Hunter and Schmidt (1990a) and other sources for additional information on individually corrected procedures.

INTERACTIVE VERSUS NONINTERACTIVE ARTIFACT DISTRIBUTION-BASED CORRECTIONS

Meta-analyses that are based on artifact distribution corrections may be performed in one of two ways—either interactively (nonlinear) or noninteractively. The term *interactive* refers to the relation between range restriction and reliability (usually predictor reliability). Just as observed correlations are attenuated by range restriction, so too are reliabilities. The more range-restricted a sample is, the more the reliability will be reduced and vice versa. Interactive procedures take this into account and correct correlations for the two artifacts (range restriction and unreliability) simultaneously. This correction is nonadditive. Alternately, *noninteractive* procedures simply assume little or no interaction between artifacts and allow the cumulative, sequential correction of artifactual sources. Interactive and noninteractive procedures yield identical results when either reliabilities or standard deviation ratios equal 1.00, and yield very similar results when the values are close to 1.00.

The procedures, codes, and examples presented in this chapter are all interactive (i.e., simultaneous correction of attenuating artifacts). Again, interested readers can refer to Hunter and Schmidt (1990a) for additional information on interactive and noninteractive procedures. Given the flexibility of the PROC MEANS procedure, the SAS code presented here can also be readily modified to perform noninteractive procedures.

CALCULATION OF SUMMARY STATISTICS

The two summary statistics that must first be calculated in a meta-analysis of correlations are the mean and the variance of the study coefficients, denoted as $\bar{r}$ and

Var(r), respectively. Examples of some commonly used formulas for converting various test statistics to *r*s are presented in Appendix C. Rosnow et al. (2000) also present a number of formulas and procedures for converting standardized effect sizes to *r*s. In computing the summary statistics, $\bar{r}$ and *Var(r)*, they are sample-weighted so that studies based on larger sample sizes are given more weight than those based on smaller samples. The conceptual basis for sample-weighting is that larger sample studies have less sample error than smaller sample studies, and thus should be assigned more weight in the computation of the mean effect. As such, the mean effect becomes a reasonable estimate of the true strength of the effect in the population. The equations for calculating the two summary statistics are presented next as Equation 3.1 and Equation 3.2 (Hunter & Schmidt, 1990a, p. 100), where N_i is the sample size for each study and r_i is the observed correlation.

$$\bar{r} = \frac{\sum(N_i * r_i)}{\sum N_i} \tag{3.1}$$

$$Var(r) = \frac{\sum(N_i * (r_i - \bar{r})^2)}{\sum N_i} \tag{3.2}$$

The SAS statements necessary to calculate the sample-size-weighted mean and the sample-size-weighted variance are presented in Table 3.1. The reader should note that all the SAS code and associated examples presented here were ran on a mainframe. A hypothetical meta-analysis of ten study correlations is shown in order to illustrate the procedures and to allow readers the chance to practice with known values. Let us assume that these ten studies represent primary studies that have investigated the relation between performance on a mental rotation test (predictor) and performance in a flight simulator (criterion). Five of the studies used only female participants and the other five used only males. The results of the meta-analysis, along with the line number of program codes producing specified results, are presented in Table 3.3.

The first *PROC MEANS* (program code line 017) calculates the sample-size-weighted mean and sample-size-weighted variance for the correlation coefficients. The second *PROC MEANS* (program code line 023) calculates the average sample size. Running this program with the hypothetical data results in a mean correlation of 0.3011921 (min = 0.02, max = 0.60, range = 0.58) and a variance across correlations of 0.0287443, both sample-size-weighted, and an average sample size of 75 (min = 25, max = 125, range = 100).

As Cohen (1990, 1994) quite eloquently discussed, it is rarely necessary to present results past two decimal places. However, to not confuse the reader, we have chosen to report the exact numbers obtained from the SAS printout. This is obviously with a clear understanding that when reporting the results in a research paper or technical report, one would typically round off most of the meta-analytic results to two decimal places (for examples, see Tables 3.7 and 3.12 in this chapter).

ESTIMATION OF SAMPLING ERROR VARIANCE

Hunter and Schmidt (1990a) noted that in most cases the largest source of the variability across study coefficients is sampling error. The effect of sampling error is that study coefficients, or validities, will vary randomly from the population value. The Hunter and Schmidt meta-analysis procedure allows removal of the variability across study coefficients that can be attributed to sampling error, resulting in a more accurate estimate of the true variance across the observed study coefficients.

Sampling error variance can be calculated using Equation 3.3 (Hunter & Schmidt, 1990a, p. 108), where $\bar{r}$ is the sample-size-weighted mean correlation calculated according to Equation 3.1, and $\bar{N}$ is the average sample size. This equation is presented next, along with corresponding calculations for this hypothetical example:

$$Var(e) = \frac{(1 - \bar{r}^2)^2}{(\bar{N} - 1)} \tag{3.3}$$

TABLE 3.1
SAS Statements for Conducting a Meta-Analysis of Correlations
With Correction for Sampling Error

```
001   DATA OBSERVED;                        ***          COMMENTS                    **;
002     INPUT STUDY R N SEX $;              *** STUDY = STUDY NUMBER, R = CORRELATION **;
003     CARDS;                              *** N = CORRESPONDING SAMPLE SIZE         **;
004   1 .60 85  FEMALE
005   2 .12 70  MALE
006   3 .24 125 MALE
007   4 .02 25  MALE
008   5 .50 75  FEMALE
009   6 .35 100 FEMALE
010   7 .16 75  MALE
011   8 .28 125 FEMALE
012   9 .07 50  MALE
013   10 .58 25 FEMALE
014   ;
015   DATA SMP_WGT1;
016     SET OBSERVED;
017   PROC MEANS MEAN VAR STD MIN MAX RANGE VARDEF = WEIGHT;
018     VAR R;
019     WEIGHT N;
020     OUTPUT OUT = ADATA MEAN = SWMC VAR= VAR_R; ***********************************;
021   DATA SMP_WGT2;                        *** SWMC = SAMPLE-WEIGHTED MEAN R **;
022     SET OBSERVED;                       ** MEAN_N = AVERAGE N ACROSS       **;
023   PROC MEANS MEAN SUM N MIN MAX RANGE;  *** ALL STUDIES, N = TOTAL N       **;
024     VAR N;                              *** K = NUMBER OF STUDIES          **;
025     OUTPUT OUT = BDATA MEAN = MEAN_N SUM = N N= K; ************************;
```

Note. The input variable R is the r (correlation coefficient) for each study and the variable N is the corresponding sample size. Most SAS users read in their data from an external data file. However, for the sake of clarity of presentation, we have chosen to present and use the data as part of the SAS program file. Finally, depending on the computer system, it may be necessary to place certain job control language (JCL) statements before the SAS statements presented in Table 3.1.

For this hypothetical example, sampling error variance is calculated as follows:

$$Var(e) = \frac{(1 - 0.3011921^2)^2}{(75.5 - 1)} \tag{3.3.1}$$

$$Var(e) = 0.011098 \tag{3.3.2}$$

This result is generated by program code lines 055 and 073-076 in Table 3.2. Removing the variance attributable to sampling error from the total variance across coefficients is a simple matter of subtracting *Var(e)* from the total variance across coefficients, *Var(r)*. For the hypothetical example, *Var(r)* is 0.0287443 and *Var(e)* is 0.011098, and the residual variance (*VAR_RES1*) is 0.017646 (i.e., 0.0287443 - 0.011098).

Note that the mean sample-weighted correlation (*SWMC*) obtained from program code lines 017-020 in Table 3.1 must be inserted in line 028 in Table 3.2. Thus, the code presented in Table 3.1 must be run first, and then the *SWMC* value inserted in line 028 in Table 3.2 and the program rerun.

Thus, given no other information about artifacts, the best estimate of the relationship between the study variables in the hypothetical example—mental rotation and performance in a flight simulator—is an average correlation of .3011921 and a residual variance of 0.017646 (*SD* = 0.1328393). Program code lines 017-020 in Table 3.1 also produce this result. The residual variance (*VAR_RES1*) is generated by program code lines 058 and 073-076 in Table 3.2. However, if other statistical artifacts are present, these estimates may not be representative of the true mental rotation ability–flight simulator performance relationship in the population.

The degree to which sampling error increases variability can be expressed by computing the percent of the observed variance that is accounted for by sampling error. Hunter and Schmidt (1990a) suggested that the presence of moderator variables can largely be ruled out if 75% or more of the variance in study coefficients can be accounted for by artifacts such as sampling error. This calculation is shown in Equation 3.4 for the hypothetical example, where the percent of variance accounted for by sampling error is denoted as PVA_{SE}.

$$PVA_{SE} = \left[\frac{Var(e)}{Var(r)}\right]*100 \qquad (3.4)$$

$$PVA_{SE} = \left[\frac{0.011098}{0.0287443}\right]*100 \qquad (3.4.1)$$

TABLE 3.2
SAS Statements for Forming Artifact Distributions, Computing the Compound Attenuation and Compound Variance Factors, and Other Test Statistics for the Meta-Analysis of Correlations

```
026   DATA ARTIFACT;
027     INPUT RXX RYY U;
028   SWMC=.3011921;
029       A = SQRT (RXX);
030       B = SQRT (RYY);
031       C = SQRT ((1 - (U ** 2)) * (SWMC ** 2) + (U ** 2));
032     CARDS;
033   .88 .60 .78
034   .74 .44 .84
035   .82 .58 .70
036   .86 .  .62
037   .78 .  .
038   ;
039   PROC MEANS MEAN VAR N MIN MAX RANGE VARDEF=N;
040       VAR A B C  RXX RYY U;
041       OUTPUT OUT=AFDATA  MEAN=MA MB MC  VAR=VA VB VC;
042
043   DATA COMPUTE;
044     MERGE ADATA BDATA AFDATA;
045       SCVA = VA / (MA * MA);
046       SCVB = VB / (MB * MB);
047       SCVC = VC / (MC * MC);
048       AA = MA * MB * MC;
049       V = SCVA + SCVB + SCVC;
050   PROC PRINT;
051       VAR SCVA SCVB SCVC AA V;
052
053   DATA FINALE;
054       SET COMPUTE;
055       VAR_E = ((1 - (SWMC ** 2)) ** 2) / (MEAN_N - 1);
056       RHO = SWMC / AA;
057       VAR_AV = (RHO ** 2) * (AA ** 2) * V;
058       VAR_RES1 = VAR_R - VAR_E;
059       VAR_RES2 = VAR_R - VAR_E - VAR_AV;
060       VAR_RHO = VAR_RES2 / (AA ** 2);
061     IF VAR_RHO LT 0 THEN VAR_RHO = 0;
062       SD_RHO = SQRT (VAR_RHO);
063       PVA_SE = 100 * (VAR_E / VAR_R);
064       PVA_ALL = 100 * (VAR_AV + VAR_E) / VAR_R;
065       SE_R1 = (1 - (SWMC ** 2)) / SQRT (N - K);
066       SE_R2 = SQRT (((((1 - (SWMC**2)) ** 2) / (N - K)) + (VAR_RES1/K));
067       L95_CONF = SWMC - (1.96 * SE_R1);
068       U95_CONF = SWMC + (1.96 * SE_R1);
069       L95_CRED = RHO - (1.96 * SD_RHO);
070       U95_CRED = RHO + (1.96 * SD_RHO);
071       CHI_SQ_S = (N / ((1 - (SWMC ** 2)) ** 2)) * VAR_R;
072       CHI_SQ = (K * VAR_R) / (VAR_AV + VAR_E);
073   PROC PRINT;
074       VAR  RHO  VAR_RHO  SD_RHO  VAR_AV  VAR_E  VAR_RES1  VAR_RES2  PVA_SE
075             PVA_ALL  SE_R1  SE_R2  L95_CONF  U95_CONF  L95_CRED  U95_CRED
076             CHI_SQ_S  CHI_SQ;
```

Note. In SAS, underscores must be used in variable names rather than a hyphen (e.g., *VAR_R* vs. *VAR-R*) to avoid confusion with the mathematical operator for subtraction.

(Table note continued)

The mean sample-weighted correlation (*SWMC*) obtained from program code lines 017-020 in Table 3.1 must be inserted in line 028 in Table 3.2. Thus, the code presented in Table 3.1 must be run first, and then the *SWMC* value inserted in line 028 in Table 3.2 and the program rerun.

RXX are the predictor unreliability values, *RYY* are the criterion unreliability values, *U* are the range restriction ratios, *AA* is the compound attenuation factor, and *V* is the compound variance factor. For the predictor unreliability distribution, *MA* is the mean, *VA* is the variance, and *SCVA* is the squared coefficient of variation; variables for criterion unreliability and range restriction are coded similarly. Periods must be inserted for any missing artifact values. The *VARDEF = N* option results in the sum of the squared deviations being divided by the number of artifact values, which is what Hunter and Schmidt (1990a) did in their illustrations, rather than the number of artifact values minus one. *VAR_R* is the sample-size-weighted variance calculated with the SAS statements in Table 3.1, and *VAR_E* is the variance attributable to sampling error. Additionally, *RHO* is the population correlation, *VAR_AV* is the variance in the observed coefficients that is attributable to study-to-study variations in the levels of the attenuating artifacts, *VAR_RES1* is the residual observed variance after removal of variance attributable to sampling error only, and *VAR_RES2* is the residual observed variance after removal of variance attributable to sampling error and study-to-study variations in attenuating artifacts. *VAR_RHO* is the variance of the fully corrected correlations in the population, and *PVA_ALL* is the percent of observed variance accounted for by sampling error and study-to-study variations in attenuating artifacts. *PVA_SE* is the percent of variance accounted for by sampling error. *SE_R1* and *SE_R2* are the standard error of the mean correlation for homogeneous and heterogeneous studies respectively. The choice of which of these two is most appropriate must be made by the user for their specific analyses. *L95_CONF* and *U95_CONF* are the lower and upper bounds of the 95% confidence interval. The 1.96 should be replaced with the appropriate *z*-value for the specified interval (i.e., 99% = 2.58; 90% = 1.64). *L95_CRED* and *U95_CRED* are the lower and upper bounds of the 95% credibility interval. Again, the 1.96 should be replaced with the appropriate *z*-value for the specified interval (i.e., 99% = 2.58; 90% = 1.64). *CHI_SQ* is the calculated chi-square test statistic ($df = K$-1) for the homogeneity of the correlations (i.e., is the unexplained variance significantly greater than zero?). The user must refer to a chi-square table to determine the significance of this test at the specified *p* value. A chi-square table is presented in Appendix D. *CHI_SQ_S* is the calculated chi-square test statistic ($df = k$-1) for correcting for sampling error only (i.e., after computing the sample-weighted correlation).

$$PVA_{SE} = 38.6092 \qquad \text{(3.4.2)}$$

This result is also produced by program code lines 063 and 073-076 in Table 3.2.

CHI-SQUARE TEST FOR VARIATION ACROSS STUDIES (DATA POINTS)

After computing the sample-weighted-mean-correlation, one can test to see whether the observed variance in the data points (or studies) is greater than that expected by chance. This is done using a chi-square test with k-1 degrees of freedom where k is the number of data points. If the chi-square test is not significant (i.e., fails to reject the null hypothesis), then there is reason to believe that there is no true variation across the data points. This chi-square test is, in essence, a test for homogeneity and can be used to assess and test for the effects of moderator variables. Specifically, this test assess whether the residual variance is significantly different from zero. Thus, if significant, it suggests the operation of possible moderator variables; that is, there may be more than one population involved and possible subpopulations should be investigated.

However, it should be noted that Hunter and Schmidt (1990a) "do not endorse this significance test because it asks the wrong question. Significant variation may be trivial in magnitude, and even nontrivial variation may still be due to research artifacts" (p. 110). They also go on to caution that "if the meta-analysis has very many studies [data points], it has very high statistical power and will therefore reject the null hypothesis, given even a trivial amount of variation across studies" (p. 112).

Nevertheless, the chi-square test statistic is calculated as follows (Hunter & Schmidt, 1990a, p. 112):

$$\chi^2_{(k-1)} = \left[\frac{N}{(1-\bar{r}^2)^2} \right] * Var(r) \tag{3.5}$$

where k is number of correlations (data points), N is the total sample size, $\bar{r}$ is the sample-size-weighted mean correlation (see Equation 3.1), and $Var(r)$ is the variance of the sample-weighted rs (Equation 3.2; program code lines 017-020 in Table 3.1). In our illustrative example, the chi-square statistic is calculated as:

$$\chi^2_{(10-1)} = \left[\frac{755}{(1-(0.3011921^2))^2} \right] * 0.0287443 \qquad (3.5.1)$$

$$\chi^2_{(9)} = 26.2482 \qquad (3.5.2)$$

This result is also produced by program code lines 071 and 073-076 in Table 3.2 (*CHI_SQ_S*). The chi-square test is significant at $p = 0.05$; the tabled chi-square value is 15.51. Thus, in this particular example, we would conclude that the data points are not homogeneous, suggesting the possible operation of moderator variables.

CONFIDENCE INTERVALS

Confidence intervals are used to assess the accuracy of the estimate of the mean correlation or effect size (i.e., $\bar{r}$ or $\bar{d}$; Whitener, 1990). *Confidence intervals* estimate the extent that sampling error remains in the sample-size-weighted mean correlation. A confidence interval gives the range of values that the mean effect size is likely to take if other sets of studies were taken from the population and used in the meta-analysis. The confidence interval is generated using the standard error for the mean correlation and is applied to the mean sample-weighted r before the correction of attenuating artifacts. In meta-analysis, a distinction is drawn between *homogeneous* and *heterogeneous* studies or data points. A homogeneous set of data points can be said to represent a single population (i.e., in meta-analytic terms, there are no moderators operating). In contrast, a heterogeneous set of data points represents multiple populations such that these data points can be classified or sorted into sublevels of a specified variable (moderator) resulting in smaller subgroups of homogeneous data points. The issue of testing for moderators in the meta-analysis of correlations is discussed later in this chapter.

The equation for calculating the standard error of the mean correlation for homogeneous studies (denoted as *SE_R1* in the SAS program code line 065 in Table 3.2) is (Whitener, 1990, p. 316):

$$SE_{r1} = \left[\frac{1 - \bar{r}^{-2}}{\sqrt{N-k}} \right] \tag{3.6}$$

The sample-size-weighted mean correlation is $\bar{r}$ (see Equation 3.1), N is the total sample size and k is the number of data points (correlations). In our specific example,

$$SE_{r1} = \left[\frac{1 - (0.3011921^2)}{\sqrt{755-10}} \right] \tag{3.6.1}$$

$$SE_{r1} = 0.033314 \tag{3.6.2}$$

This result is also produced by program code lines 065 and 073-076 in Table 3.2. A 95% confidence interval is subsequently calculated as follows:

$$Conf_{95\%} = \bar{r} \pm (1.96) * (SE_{r1}) \tag{3.7}$$

that in our specific example is:

$$Conf_{95\%} = 0.3011921 \pm (1.96) * (0.033314) \tag{3.7.1}$$

Thus, the 95% confidence interval for our sample-weighted mean correlation is:

$$0.23590 < \bar{r} < 0.36649 \tag{3.7.2}$$

These results are produced by program code lines 067-068 and 073-076 in Table 3.2. The 1.96 should be replaced with the appropriate z-value for the specified interval (i.e., 99% = 2.58; 90% = 1.64).

Because the confidence interval reflects the extent of sampling error that remains in the sample-size-weighted mean r, the interval should be calculated before the correlation is corrected for statistical artifacts. A relatively small confidence interval band that does not include zero is considered a favorable outcome. The results from our example indicate that relatively little sampling error remains in the sample-size-weighted mean correlation.

However, in this hypothetical example, because the results of the chi-square test for variation indicated our data points are not homogeneous, the equation for heterogenous studies is more appropriate for calculating the standard error of the mean correlation (denoted as SE_R2 in the SAS program code line 066 in Table 3.2). This equation is presented below (Whitener, 1990, p. 317),

$$SE_{r2} = \sqrt{\left[\frac{(1-\bar{r}^2)^2}{N-k}\right] + \left[\frac{Var_{res}}{k}\right]} \tag{3.8}$$

where $\bar{r}$ is the sample-size-weighted mean correlation (see Equation 3.1), N is the total sample size, k is the number of data points (correlations), and Var_{res} is the residual variance, that is, the variance in the observed correlations after the variance due to sampling error has been removed. Var_{res} is denoted as VAR_RES1 in the SAS program code (see lines 058 and 073-076 in Table 3.2). In our specific example,

$$SE_{r2} = \sqrt{\left[\frac{(1 - (0.3011921^2)^2}{755 - 10}\right] + \left[\frac{0.017646}{10}\right]} \qquad (3.8.1)$$

$$SE_{r2} = 0.053614 \qquad (3.8.2)$$

This result is also produced by program code lines 066 and 073-076 in Table 3.2. A 95% confidence interval is subsequently calculated as follows:

$$Conf_{95\%} = \bar{r} \pm (1.96) * (SE_{r2}) \qquad (3.9)$$

that in our specific example is:

$$Conf_{95\%} = 0.3011921 \pm (1.96) * (0.053614) \qquad (3.9.1)$$

Thus, the 95% confidence interval for our sample-weighted mean correlation—based on the standard error of the mean correlation for heterogeneous studies—is:

$$0.1961087 < \bar{r} < 0.4062755 \qquad (3.9.2)$$

The program code for Equation 3.9 is not presented in Table 3.2. However, this can be accomplished by replacing *SE_R1* in lines 067 and 068 with *SE_R2*. The 1.96 should also be replaced with the appropriate z-value for the specified interval (i.e., 99% = 2.58; 90% = 1.64).

In summary, although we have presented computations and results for both the standard error of the mean correlation for homogeneous (SE_R1) and heterogeneous (SE_R2) studies in the preceding section, the decision of which is more appropriate must be made by the researcher. This decision is guided by the chi-square test for variation across the correlations (see Equation 3.5). In our specific hypothetical example, because the chi-square test was significant and suggested that the correlations are not homogenous, the appropriate choice is the standard error of the mean correlation for heterogeneous studies.

The SAS program code and the results of the hypothetical test validity example are presented in Tables 3.1, 3.2, and 3.3, respectively.

CORRECTING SUMMARY STATISTICS FOR MULTIPLE ATTENUATING ARTIFACTS

Although sampling error normally affects only the variance across study coefficients, other artifacts such as measurement error, range restriction, and the dichotomization of continuous variables affect both the mean and the variance of the study coefficients (Hunter & Schmidt, 1990a). These artifacts are referred to as "attenuating artifacts" since they tend to systematically reduce the mean correlation. (See Hunter and Schmidt, 1990a, for a detailed discussion of attenuating artifacts.)

As previously noted, Steiner et al.'s (1991) survey of the organizational behavior/human resource management literature indicated that corrections are frequently made for attenuating artifacts. In particular, while all of the Hunter and Schmidt-based meta-analytic studies surveyed corrected for sampling error, 69.2% also corrected for criterion unreliability, 50% corrected for predictor unreliability, and 11.5% corrected for range restriction in the predictor variable. We next present the SAS PROC MEANS procedures for correcting for these three common attenuating artifacts.

TABLE 3.3
**Meta-Analysis of Correlations Results for Hypothetical Mental Rotation Test
Validity Data**

Meta-Analysis Statistic	Results	Program Code Line Numbers from Tables 3.1 and 3.2
Sampling Error Results		
Total N	755	023-025
Mean N	75	023-025
Min N	25	023-025
Max N	125	023-025
Range of N	100	023-025
Number of rs	10	023-025
Mean observed r (sample-weighted)	0.3011921	017-020
Min observed r	0.02	017-020
Max observed r	0.60	017-020
Range of observed rs	0.58	017-020
Variance of observed rs	0.0287443	017-020
SD of observed rs	0.1695414	017-020
Variance due to sampling error	0.011098	055; 073-076
Percent variance accounted for	36.6092	063; 073-076
Residual variance	0.017646	058; 073-076
Chi-square value	26.2482	071; 073-076

(continued)

95% Confidence Interval

Lower limit	0.23590	067; 073-076
Upper limit	0.36649	068; 073-076

Attenuation Artifact
Corrections

Mean true r (ρ)	0.59594	056; 073-076
Var of true r (ρ)	0.063901	060; 073-076
SD of true r (ρ)	0.25279	062; 073-076
Percent variance due all artifacts	43.2136	064; 073-076
Residual variance	0.016323	059; 073-076

95% Credibility Interval

Lower limit	0.10047	069; 073-076
Upper limit	1.09140	070; 073-076

Chi-square value	23.1408	072-076

Artifact Values

Predictor reliability
N	5	
Mean	0.816	039-041

Criterion reliability
N	3	
Mean	0.54	039-041

Restricted SD (Range
restriction)
N	4	039-041
Mean	0.735	

(continued)

Mean of square root of predictor reliability	0.9028784	029; 039-041
Mean of square root of criterion reliability	0.7331663	030; 039-041
Range restriction attenuation factor	0.7635024	031; 039-041

Artifact Distributions

Although the reporting of this information in primary studies is somewhat spo-
radic, correcting the summary statistics for attenuating artifacts first requires the
compilation of a list of the actual artifact values obtained from the primary studies. The values in each distribution of attenuating artifacts are then converted to
the appropriate psychometric form. Measurement error is converted by taking the
square root of the reliability coefficients (Ghiselli, Campbell, & Zedeck, 1981).
Range restriction data is first converted to u, the ratio of the standard deviation of
the restricted sample to the standard deviation of the unrestricted sample, and then
entered into the range restriction formula (Callender & Osburn, 1980). The equations are shown below for measurement error in the predictor (r_{xx}), measurement
error in the criterion (r_{yy}), and range restriction (u). Following the convention of
Hunter and Schmidt (1990a), these attenuation factors are expressed sequentially
as lowercase letters starting with a.

$$a = \sqrt{r_{xx}} \tag{3.10}$$

$$b = \sqrt{r_{yy}} \tag{3.11}$$

$$c = \sqrt{\left[(1-u^2)*(\overline{r}^{-2})+(u^2)\right]} \qquad (3.12)$$

After conversion, three values are extracted from each artifact distribution: the mean of the distribution, the corresponding variance, and the squared coefficient of variation. Both the mean and the variance are calculated in the usual manner and are not sample-size weighted. The squared coefficient of variation is computed by dividing the variance of the artifact distribution by the square of the mean.

To illustrate, assume that the following artifact values were collected from the studies in the hypothetical mental rotation test/simulator flight performance example. Reliability coefficients for the predictor were reported in five of the ten hypothetical studies. These values were 0.88, 0.74, 0.82, 0.86, and 0.78, respectively. Criterion reliability coefficients were reported in three of the studies, 0.60, 0.44, and 0.58. Finally, range restriction ratios were reported in four of the studies, 0.78, 0.84, 0.70, and 0.62. The necessary sequence of SAS statements for converting these artifactual values and computing the mean, variance, and squared coefficient of variation of each distribution are shown in Table 3.2, program code lines 029-041.

Running this program results in a mean of 0.9028784, denoted as $\overline{a}$, and a variance of 0.000810548, denoted as $Var(a)$, for the artifact predictor reliability. For criterion reliability, the result is a mean of 0.7331663, denoted as $\overline{b}$, and a variance of 0.0024672, denoted as $Var(b)$. For range restriction, the result is a mean of 0.7635025, denoted as $\overline{c}$, and a variance of 0.0052496, denoted as $Var(c)$. These results are produced by program code lines 029-031, and 039-041. The squared coefficients of variation are 0.00099431, 0.0045898, and 0.0090054, respectively (see program code lines 039-051 in Table 3.2).

Compound Attenuation Factor

The three values extracted from the artifact distributions are used to compute two correction factors; these are the compound attenuation factor (AA—Equation 3.13) and the compound variance factor (V—Equation 3.15). The compound attenuation factor, denoted as AA (in the SAS program code), is formed by multiplying the average values of the attenuating artifact distributions together (Hunter & Schmidt, 1990a). The equation for computing the compound attenuation factor is presented in Equation 3.13 along with the calculations for the hypothetical example. The SAS statement for calculating the compound attenuation factor is also presented in Table 3.2 in the *DATA COMPUTE* section, program code lines 048 and 050-051.

$$AA = \bar{a} * \bar{b} * \bar{c} \tag{3.13}$$

$$AA = 0.9028784 * 0.7331663 * 0.7635024 \tag{3.13.1}$$

$$AA = 0.50541 \tag{3.13.2}$$

ρ (rho)—Estimate of Fully Corrected Population Correlation

As noted previously, one effect of attenuating artifacts is to reduce the average correlation between the study variables. To correct for this decrement, the mean observed correlation is divided by the compound attenuation factor (AA), which corrects the mean observed correlation for the average effects of the attenuating artifacts. The resulting value represents an estimate of the fully corrected correlation in the population (ρ). The estimate of rho is 0.59594 for the hypothetical example; this calculation is shown in Equation 3.14. The SAS statements for computing ρ are presented in Table 3.2, program code lines 056 and 073-076.

$$\rho = \frac{\bar{r}}{AA} \tag{3.14}$$

$$\rho = \frac{0.3011921}{0.50541} \tag{3.14.1}$$

$$\rho = 0.59594 \qquad\qquad (3.14.2)$$

Compound Variance Factor

The compound variance factor, denoted as V, is formed by summing the individual squared coefficients of variation for each attenuating artifact distribution. Computation of the compound variance factor for the hypothetical example is shown in Equation 3.15, where each lowercase scv represents a squared coefficient of variation. The SAS statements for calculating the compound variance factor are also included in the *DATA COMPUTE* section of Table 3.2, program code lines 045-047 and 049-051.

$$V = scv(a) + scv(b) + scv(c) \qquad\qquad (3.15)$$

$$V = 0.00099431 + 0.0045898 + 0.0090054 \qquad\qquad (3.15.1)$$

$$V = 0.014590 \qquad\qquad (3.15.2)$$

Variance due to Attenuating Artifacts

The presence of attenuating artifacts also affects the variance of the study correlations in two ways. First, study to study variations in the levels of the attenuating artifacts artificially increase the variability across the observed correlations over and above the effects of sampling error. The amount of variance in the observed correlations which is attributable to study to study variations in the levels of the attenuating artifacts, denoted as $Var(AV)$, can be computed using the equation provided by Hunter and Schmidt (1990a, p. 179). This equation is presented below in Equation 3.16, along with the corresponding calculations for the hypothetical example. The SAS statements for computing $Var(AV)$ are presented in Table 3.2, program code lines 057 and 073-076.

$$Var(AV) = \rho^2 * AA^2 * V \qquad (3.16)$$

$$Var(AV) = 0.59594^2 * 0.50541^2 * 0.014590 \qquad (3.16.1)$$

$$Var(AV) = 0.0013235 \qquad (3.16.2)$$

True (Residual) Variance in *rs*

Removing the variance in the observed correlations that can be attributed to both sampling error and study-to-study artifact variations results in an estimate of the true (residual) variance in the observed coefficients, which can be denoted as *Var(RES$_T$)*. The equation for computing the residual observed variance is presented in Equation 3.17, along with the corresponding calculations for the hypothetical example. The SAS statements for this statistic are also presented in Table 3.2, program code lines 059 and 073-076. In the SAS code, *Var(RES$_T$)* is denoted as *VAR_RES2*.

$$Var(RES_T) = Var(r) - Var(e) - Var(AV) \qquad (3.17)$$

$$Var(RES_T) = 0.0287443 - 0.011098 - 0.0013235 \qquad (3.17.1)$$

$$Var(RES_T) = 0.016323 \qquad (3.17.2)$$

Total Variance Accounted for by Sampling Error and Attenuating Artifact Factors

The calculation for determining the total percent of the observed variance accounted for by sampling error and attenuating artifacts, denoted as *PVA_ALL*, is shown in Equation 3.18. The SAS statements for this statistic are presented in Table 3.2, program code lines 064 and 073-076.

$$PVA\_ALL = \left[\frac{Var(e) + Var(AV)}{Var(r)} \right] *100 \qquad (3.18)$$

$$PVA\_ALL = \left[\frac{0.011098 + 0.0013235}{0.0287443} \right] *100 \qquad (3.18.1)$$

$$PVA\_ALL = 43.2136 \qquad (3.18.2)$$

Variance of ρ (rho)

Attenuating artifacts also serve to make the residual observed variance an underestimate of the corresponding residual (true) variance in the population (see Hunter and Schmidt, 1990a). For example, if all of the studies are range restricted, then there will not be as much variability across them as there would have been if there was no range restriction. Essentially, it is because the larger effect sizes are reduced more by range restriction than the smaller effect sizes, thereby making them all closer in magnitude. An estimate of the variance of the fully corrected correlations in the population (denoted as $Var(\rho)$ or VAR_RHO in the SAS code) can be obtained by adjusting the residual observed variance based on the average levels of the artifact distributions. Specifically, the residual observed variance $Var(RES_\rho)$ is divided by the square of the compound attenuation factor. This equation, along with the corresponding calculation for the hypothetical example is shown in Equation 3.19. The SAS statements for this statistic are presented in Table 3.2, program code lines 060 and 073-076.

$$Var(RHO) = \frac{Var(RES_T)}{AA^2} \qquad (3.19)$$

$$Var(RHO) = \frac{0.016323}{0.50541^2} \qquad (3.19.1)$$

$$Var(RHO) = 0.063901 \qquad (3.19.2)$$

Note that program code line 061 sets VAR_RHO to zero (0) if it is less than zero (i.e., a negative value). This is necessary to compute the standard deviation of $Var(\rho)$ (the square-root of $Var(\rho)$—see Equation 3.20). That is, if $Var(\rho)$ is negative, its square-root will be an invalid mathematical operation.

In summary, the preceding computations yield two new summary statistics that describe the true relation between the study variables in the population: the average corrected correlation (rho), and the corresponding variance of rho in the population. Thus, in our hypothetical example, the best estimate of the relation between mental rotation test performance and flight simulator performance in the population is a correlation of 0.59594 with a variance of 0.063901 ($SD = 0.25279$).

It should be emphasized that the procedures described in the preceding section are not necessarily limited to the three attenuating artifacts described here. We limited our discussion to these three because they are the most frequently made corrections (Steiner et al., 1991). In fact, these procedures can be used to correct for any number of attenuating artifacts (and any combination thereof) simply by including more or fewer terms in the respective equations. This feature allows a high degree of flexibility in adapting the meta-analysis to the information provided in the primary studies. The only requirement for correcting a particular attenuating artifact is that an artifact distribution can be formed. Specific conversion formulas can be obtained from Hunter and Schmidt (1990a). Also see Hunter and Schmidt (1990b) for an in-depth discussion of the dichotomization of continuous variables, its effect on and implications for meta-analysis, and the necessary corrections.

CONDUCTING MODERATOR ANALYSIS IN THE META-ANALYSIS OF CORRELATIONS

In general, variable Z is considered to be a moderator of the relation between variables X and Y when the nature of this relation is contingent on the levels or values of Z. In the context of meta-analysis, a moderator variable is defined as any variable that by its inclusion in the analysis accounts for, or helps explain, more variance than would otherwise be the case. This is somewhat different from the use of the term *moderator* in multiple regression. In multiple regression, a moderator is a variable that, although having a negligible correlation with a criterion, interacts with another variable to enhance the predictability of the criterion variable (Cohen & Cohen, 1983). In the meta-analysis of correlations, when sufficient variance remains in the population (i.e., ρ) after the specified corrections have been made, the presence of one or more moderator variables is suspected. Alternately, various moderator variables may be suggested by theory. Thus, the decision to search and/or test for moderators may be driven either theoretically or empirically. So, for instance, in our mental rotation test validity example, we may have some theoretical or conceptual basis to hypothesize that the validity of the test may be a function of the sex of the test taker—either male or female. And on this basis, we will then make a decision to test for sex as a moderator.

On the other hand, this decision could be made primarily on the basis of the amount of variance accounted for and the degree of variability in ρ after all the specified corrections have been made. Thus, in meta-analysis, a moderator variable is identified if: the variance of the average correlations is lower in the subsets or sublevels of the variable than the aggregate level, and/or the average correlation varies from subset to subset. In short, in meta-analysis, if large differences are found between sublevels of a given variable, then that variable can be considered a moderator variable.

A number of statistical tools or methods have been used to make decisions about whether any of the observed variance in the corrected correlation (ρ) across studies is real and not solely the result of artifacts. Specifically, these methods are used to search and test for the presence and effect of potential or hypothesized moderator variables. Examples of these methods include: (a) using the 75% rule, (b) testing for significant differences between the corrected mean correlations of sublevels of the hypothesized moderator variable, (c) using a chi-square test of homogeneity to determine if the unexplained residual variance is significantly different from zero or phrased differently, to test if the observed correlations are from the same population or from different subpopulations (i.e., different levels of the moderator), and (d) computing credibility intervals about the fully corrected average correlation (ρ).

The 75% rule is more commonly associated with the meta-analysis of correlations and states that if 75% or more of the variance can be accounted for by the specified corrections, then it can be concluded that all of the variance is due to artifacts because the remaining 25% is likely to be due to uncorrected artifacts such as computational and typographical errors. On the other hand, if the variation in correlations is determined not to be artifactual, but real, then moderator variables may be operating. There are generally two types of moderators: those that are theoretically predicted, and those that are unsuspected. For those moderators that have been theoretically postulated, one could use statistical significance tests comparing the mean correlations of the various levels of the moderator variable (Alliger, 1995; see also Aguinis & Pierce, 1998). For unsuspected or unpredicted moderating variables, one must examine the residual variation left over after the specified corrections have been made to determine if it is significantly different from zero. Consequently, a chi-square test can be used to try to detect these moderators by testing for the homogeneity of the true correlations across studies.

Finally, credibility intervals can also used to investigate the possibility that moderator variables that may affect the variance in the corrected average correlations.

Because these four methods answer somewhat different questions and also use different decision rules, the results obtained from them may or may not be congruent (e.g., see Arthur et al., 1991). Consequently, it is usually advisable to use at least two or more of these methods in the search and test for moderators.

As previously cautioned in chapter 2, the use of statistical significance tests to test for moderators in meta-analysis may be misguided (see Hunter & Schmidt, 1990a; Schmidt, 1992). It has been argued that the use of significance tests in primary research under conditions of low power (almost universal in research literatures) is the very reason why research literatures appear contradictory and confusing in the first place (see Cortina & Dunlap, 1997; Hagen, 1997; Schmidt, 1992, 1996; Schmidt & Hunter, 1978; Thompson, 1996; Wilkinson et al., 1999, for a discussion of this issue), and consequently, the reason why meta-analysis is *necessary* to make sense of these literatures. It is, therefore, illogical to introduce this same problem to meta-analysis. For instance, Hunter and Schmidt (1990a) do not endorse the chi-square test for homogeneity because "this significance test . . . asks the wrong question. Significant variation may be trivial in magnitude, and even nontrivial variation may still be due to research artifacts" (p. 110). They also go on to caution that "if the meta-analysis has very many studies [data points], it has very high statistical power and will therefore reject the null hypothesis, given even a trivial amount of variation across studies" (p. 112).

Credibility Intervals

Credibility intervals are used to search and test for the presence of potential or hypothesized moderator variables. Specifically, they are used to investigate the possibility that moderator variables may be affecting the variance in the corrected average correlations. If moderator variables are operating (under these circumstances, the credibility interval will include zero), there may be more than one population involved and possible subpopulations should be investigated. Because the corrected mean effect size is interpreted as an estimate of the population mean effect size, the credibility interval helps determine the generalizability of the obtained corrected mean correlations by providing information about the existence of moderator variables. Unlike confidence intervals, which are generated using the standard error of the mean correlation before statistical artifacts are corrected for, credibility intervals are generated after the mean correlation has been corrected for attenuating artifacts. The credibility interval is generated by using the corrected standard deviation around the mean of the distribution of true correlations ($SD\rho$ or SD_RHO in the SAS code), where:

$$SD\rho = \sqrt{Var(\rho)} \qquad (3.20)$$

$Var(\rho)$ is calculated using Equation 3.19 (program code lines 060 and 073-076). The program code for $SD\rho$ is presented in Table 3.2, program code lines 062 and 073-076. In our specific illustrative example, $SD\rho$ is as follows:

$$SD\rho = \sqrt{0.063901} \qquad (3.20.1)$$

$$SD\rho = 0.25279 \qquad (3.20.2)$$

A 95% credibility interval is subsequently calculated as:

$$Cred_{95\%} = \rho \pm (1.96)*(SD\rho) \qquad (3.21)$$

In our specific example, the 95% credibility interval is:

$$Cred_{95\%} = 0.59594 \pm (1.96)*(0.25279) \qquad (3.21.1)$$

$$Cred_{95\%} = 0.59594 \pm 0.4954684 \qquad (3.21.2)$$

Thus, the 95% credibility interval for our fully corrected $\bar{r}$ (ρ) is as follows:

$$0.10047 < \rho < 1.09140 \qquad (3.21.3)$$

These results are produced by program code lines 069-070 and 073-076 in Table 3.2. Like the confidence interval, the 1.96 should be replaced with the appropriate z-value for the specified interval (i.e., 99% = 2.58; 90% = 1.64).

It is interesting to note that the upper bound of ρ is greater than 1.00. This is a result of using artifact distributions for the meta-analytic corrections. Although correcting each study individually for artifacts is more methodologically precise, it is rarely possible in practice because by necessity all studies must report the relevant information. Using a sample of artifact values to correct the summary statistics computed across all studies (the artifact distribution approach) is a reasonable alternative. But, it is an approximation and as such, the artifact values used to form the distributions contain a certain degree of sampling error. In turn, this sampling error can lead to either over- or under-estimation of the resulting summary values, including the occasional instance where the endpoint of a credibility or confidence interval for a correlation estimate exceeds 1.00 or the percent-of-variance accounted for exceeds 100%. In either instance, the out-of-bound values are simply corrected to the logical endpoint. For example, a corrected correlation greater than 1.00 is rounded down to 1.00 and a percent-of-variance greater than 100% is rounded down to 100%.

Confidence intervals and credibility intervals must be used together to determine whether the corrected mean correlation is the estimate of the population parameter or is representing the mean of more than one subpopulation (Whitener, 1990). Although the credibility interval can detect the presence of moderating variables, it cannot identify which moderators are operating. Thus, a credibility interval first must be generated around the corrected mean correlation using the corrected standard deviation. The corrected mean correlation is probably estimating one population parameter without moderators operating if the credibility interval is small or does not include zero. Under these conditions, the data points can be said to be homogeneous. However, as previously noted, if the interval includes zero or is sufficiently large, then moderators are probably operating and the corrected mean correlation size is really the mean effect of several subpopulations. Thus, in our illustrative example, although the credibility interval does not include zero it is sufficiently large to suggest the presence and operation of moderators. The data points can be considered to be heterogeneous.

Confidence Intervals

Once subgroups are identified and it is confirmed that the corrected mean correlations of these subgroups are not themselves affected by the operation of still other moderating variables, confidence intervals can then be used to estimate the accuracy of the sample-size-weighted mean correlation for each subgroup. Again, the confidence intervals around each subgroup mean uses the standard error for the mean correlation for homogeneous studies (SE_{rl} or SE_R1 in the SAS program code). This is after it has been established (using credibility intervals and/or chi-square tests) that no further moderating variables are operating. As was previously presented in Equation 3.6, the equation for SE_{rl} is as follows (Whitener, 1990, p. 316):

$$SE_{r1} = \left[\frac{1 - \bar{r}^2}{\sqrt{N - k}} \right] \tag{3.22}$$

where $\bar{r}$ is the sample-size-weighted mean correlation (see Equation 3.1), N is the total sample size, and k is the number of data points (correlations).

A confidence interval can also be generated around the average of several subpopulation parameters but must use the standard error for a heterogeneous case (because the mean of subpopulations has moderators present). The equation for computing the standard error in the mean correlation for heterogenous studies (denoted as SE_R2 in the SAS program code line 066, Table 3.2) was previously presented in Equation 3.8 as follows (Whitener, 1990, p. 317):

$$SE_{r2} = \sqrt{\left[\frac{(1 - \bar{r}^2)}{N - k} \right] + \left[\frac{Var_{res}}{k} \right]} \tag{3.23}$$

where $\bar{r}$ is the sample-size-weighted mean correlation (see Equation 3.1), N is the total sample size, k is the number of data points (correlations), and Var_{res} is the residual variance, that is, the variance in the observed correlations after the variance due to sampling error has been removed. Var_{res} is denoted as VAR_RES1 in the SAS program code (see lines 058 and 073-076 in Table 3.2).

Chi-Square Test for Homogeneity of Corrected Correlations

The chi-square tests the hypothesis that the observed variance is due to sampling error and artifactual variation. Thus, the null hypothesis is that there is no real variance in the fully corrected, unattenuated correlations. Specifically, the residual variance should equal zero if no moderators are operating. The test statistic Q is calculated by Hunter and Schmidt (1990a, p. 168) as follows:

$$Q_{(k-1)} = \left[\frac{K * Var(r)}{(\rho^2 * AA^2 * V) + Var(e)} \right] \qquad (3.24)$$

$$Q_{(k-1)} = \left[\frac{k * Var(r)}{Var(AV) + Var(e)} \right] \qquad (3.24.1)$$

where k = number of correlations (data points); $Var(r)$ = variance of the sample-weighted rs (Equation 3.2, program code lines 017-020 in Table 3.1); ρ = rho (Equation 3.14, program code lines 056 and 073-076 in Table 3.2); AA = compound attenuation factor (Equation 3.11, program code lines 048 and 050-051 in Table 3.2); V = compound variance factor (Equation 3.15, program code lines 045-047 and 049-051); and $Var(e)$ = sampling error variance (Equation 3.3, program code lines 055 and 073-076 in Table 3.2). $Var(AV)$ is presented in Equation 3.16 (program code lines 057 and 073-076). Q will approximate a chi-square distribution with k-1 degrees of freedom. In the SAS program code, the Q test statistic is denoted as CHI_SQ.

On a cautionary note, although the chi-square test is presented and discussed in their book, Hunter and Schmidt (1990a) are not particularly enthusiastic about the use of this significance test. They argue that being a significance test, the chi-

square test suffers from all the problems and flaws associated with most significance tests.

Equation 3.24.1 is presented in the SAS program code in Table 3.2 in program lines 072-076 and also in our next example:

$$Q_{(k-1)} = \left[\frac{10*0.0287443}{0.0013235*0.011098} \right] \qquad (3.24.2)$$

$$Q_{(k-1)} = 23.1408 \qquad (3.24.3)$$

This chi-square value is significant at $p = .05$ (the tabled value required for
significance is 15.51). A chi-square table is presented in Appendix D of this volume. If the value obtained by the chi-square test is less than the value required for
statistical significance, the test suggests that the research domain is homogeneous
with no moderator variables present. Conversely, larger chi-square value indicates
that the unexplained variance is significantly greater than zero, suggesting that the
remaining variance is due to additional moderators or statistical artifacts. However, it is important to note that because of the low power of the chi-square test, if
the number of data points or the variation in the effect sizes caused by the moderator variables is small, the chi-square test may produce an insignificant value
when moderating variables are actually operating. The only way to overcome the
limitation of the low power of the chi-square test is to gather enough studies to
generate a large total number of data points. Yet, as noted previously, when the
meta-analysis contains a large number of data points, the high level of statistical
power increases the likelihood of rejecting the null hypothesis even in the presence of very trivial amounts of variation across studies. This quandary highlights
the reason some advocates of meta-analysis, such as Schmidt and Hunter (1990a),
are against the use of statistical significance tests in meta-analysis.

Nevertheless, after one or more of the above statistical tools (75% rule, credibility intervals, chi-square test) have been used to assess the presence of moderators, the data set is typically separated into subsets according to the various levels
of the potential moderator variable. The mean correlation and variance is then

recalculated separately for each subgroup (e.g., males vs. females). Again, the moderator variable is generally confirmed if the means of the subgroups differ and the variance within each group is less than the original variance. Confidence intervals around the sample-weighted mean correlation for each subgroup can then be calculated using the standard error for the mean correlation, SE_{rl}, for homogeneous studies (Equations 3.6 and 3.22).

Analysis of Moderator Variables

The analysis of moderator variables using the SAS PROC MEANS procedure is relatively simple. The level of the moderator variable is first coded and entered along with the r and sample size for each study (e.g., male and female test takers in Table 3.1). An *IF* statement can then be used to select studies that used only males (*IF SEX = 'MALE';*) or only females (*IF SEX = 'FEMALE';*) into the analyses. The complete meta-analysis statistics are then recalculated for each level of the moderator variable. A moderator is identified by a corrected variance that has a lower average in the subsets than for the data as a whole, and a corrected mean r that varies from subset to subset.

Alternatively, one could use *PROC SORT* coupled with *BY* or even *WHERE* statements to run these analyses. The advantage to *PROC SORT*s is that one does not have to separately rerun the analysis for each level of the moderator because analyses of all the different levels can be accomplished within a single run. The disadvantage is that although it is not difficult to master, the researcher must be familiar with *PROC SORT*s and the use multiple *PROC SORT* statements within the same program. For the sake of simplicity and clarity we have limited our presentation and examples to the use of *IF* statements.

The SAS code (using *IF* statements) necessary to run a moderator analysis of test taker sex (male vs. female) in our hypothetical mental rotation test validity example is presented in Table 3.4. The SAS printout for the moderator analysis is presented in Table 3.6. In general, a moderator is identified by (a) a corrected variance that has a lower average in the subsets than for the data as a whole, and a corrected mean r that varies from subset to subset. Consistent with the results obtained for the overall analysis and its interpretation shown in Table 3.7, the results of the moderator analysis show that the test taker's sex does operate as a

TABLE 3.4
SAS Statements for Conducting a Moderator Analysis for Mental Rotation Test Example With Test Taker Sex (Female) as a Moderator Variable for the Meta-Analysis of Correlations

```
001    DATA OBSERVED;                    ***              COMMENTS                **;
002    INPUT STUDY R N SEX $;            *** STUDY=STUDY NUMBER, R=CORRELATION    **;
003      CARDS;                          *** N = CORRESPONDING SAMPLE SIZE        **;
004    1 .60 85  FEMALE
005    2 .12 70  MALE
006    3 .24 125 MALE
007    4 .02 25  MALE
008    5 .50 75  FEMALE
009    6 .35 100 FEMALE
010    7 .16 75  MALE
011    8 .28 125 FEMALE
012    9 .07 50  MALE
013    10 .58 25 FEMALE
014    ;                                 **************************************************;
015    DATA SMP_WGT1;                    ** FOR MALES CHANGE 'FEMALE'             **;
016      SET OBSERVED;                   ** TO 'MALE'                            **;
017    IF SEX ='FEMALE';                 **************************************************;
018    PROC MEANS MEAN VAR STD MIN MAX RANGE VARDEF=WEIGHT;
019          VAR R;
020          WEIGHT N;
021          OUTPUT OUT=ADATA MEAN=SWMC VAR=VAR_R;
022    DATA SMP_WGT2;                     **************************************************;
023      SET OBSERVED;                   ** FOR MALES CHANGE 'FEMALE'             **;
024    IF SEX='FEMALE';                  ** TO 'MALE                             **;
025    PROC MEANS MEAN SUM N MIN MAX RANGE;  **********************************************;
026          VAR N;                      *** SWMC=SAMPLE-WEIGHTED MEAN R          **;
027          OUTPUT OUT=BDATA MEAN=MEAN_N SUM=N N=K;  * MEAN_N=AVERAGE N ACROSS   **;
028                                       *** ALL STUDIES, N=TOTAL N               **;
029    DATA ARTIFACT;                    *** K=NUMBER OF STUDIES                  **;
030      INPUT RXX RYY U;                **************************************************;
031    SWMC=.4219512;
032          A = SQRT (RXX);
033          B = SQRT (RYY);
034          C = SQRT ((1 - (U ** 2)) * (SWMC ** 2) + (U ** 2));
035       CARDS;
036    .88 .60 .78
037    .74 .44 .84
038    .82 .58 .70
039    .86 .  .62
040    .78 .  .
041    ;
042    PROC MEANS MEAN VAR N MIN MAX RANGE VARDEF=N;
043          VAR A B C  RXX RYY U;
044          OUTPUT OUT=AFDATA  MEAN=MA MB MC  VAR=VA VB VC;
```

(continued)

moderator of the mental rotation test ability–flight simulator performance relationship. A summary of the meta-analysis results for the mental rotation test example is presented in Table 3.7.

```
045   DATA COMPUTE;
046     MERGE ADATA BDATA AFDATA;
047        SCVA = VA / (MA * MA);
048        SCVB = VB / (MB * MB);
049        SCVC = VC / (MC * MC);
050        AA = MA * MB * MC;
051        V = SCVA + SCVB + SCVC;
052   PROC PRINT;
053        VAR SCVA SCVB SCVC AA V;
054
055   DATA FINALE;
056     SET COMPUTE;
057        VAR_E = ((1 - (SWMC ** 2)) ** 2) / (MEAN_N - 1);
058        RHO = SWMC / AA;
059        VAR_AV = (RHO ** 2) * (AA ** 2) * V;
060        VAR_RES1 = VAR_R - VAR_E;
061        VAR_RES2 = VAR_R - VAR_E - VAR_AV;
062        VAR_RHO = VAR_RES2 / (AA ** 2);
063     IF VAR_RHO LT 0 THEN VAR_RHO = 0;
064        SD_RHO = SQRT (VAR_RHO);
065        PVA_SE = 100 * (VAR_E / VAR_R);
066        PVA_ALL = 100 * (VAR_AV + VAR_E) / VAR_R;
067        SE_R1 = (1 - (SWMC ** 2)) / SQRT (N - K);
068        SE_R2 = SQRT ((((1 - (SWMC**2)) ** 2) / (N - K)) + (VAR_RES1/K));
069        L95_CONF = SWMC - (1.96 * SE_R1);
070        U95_CONF = SWMC + (1.96 * SE_R1);
071        L95_CRED = RHO - (1.96 * SD_RHO);
072        U95_CRED = RHO + (1.96 * SD_RHO);
073        CHI_SQ_S = (N / ((1 - (SWMC ** 2)) ** 2)) * VAR_R;
074        CHI_SQ = (K * VAR_R) / (VAR_AV + VAR_E);
075   PROC PRINT;
076        VAR  RHO VAR_RHO SD_RHO VAR_AV VAR_E VAR_RES1 VAR_RES2 PVA_SE
077             PVA_ALL SE_R1 SE_R2 L95_CONF U95_CONF L95_CRED U95_CRED
078             CHI_SQ_S CHI_SQ;
```

Note. The above SAS code is for females only. To run the meta-analysis for males, the level of the moderator variable will have to be selected in the *IF* statements and the program rerun. (The SAS code for all the moderators use in this example can be found at www.erlbaum.com.) An alternative is to run the analyses for both levels of the moderator in the same program by repeating or cutting and pasting the males' code to the end of the females' code. Running all levels of the moderator in the same program can also be accomplished by using *PROC SORT* or *WHERE* statements.

In SAS, underscores must be used in variable names rather than hyphens (e.g., *VAR_R* versus *VAR-R*) to avoid confusion with the mathematical operator for subtraction.

The mean sample-weighted correlation—*SWMC*—obtained from program code lines 018-021 must be inserted in line 031. Thus, the code presented in Table 3.4 must be run first, and then the *SWMC* value inserted in line 031 and the program rerun.

The input variable *R* is the *r* (correlation coefficient) for each study and the variable *N* is the corresponding sample size. Most SAS users read in their data from an external data file. However, for the sake of clarity of presentation, we have chosen to present and use the data as part of the SAS program file. Finally, depending on the computer system, it may be necessary to place certain job control language (JCL) statements before the SAS statements presented in Table 3.4.

(Table note continued)

RXX are the predictor unreliability values, *RYY* are the criterion unreliability values, *U* are the range restriction ratios, *AA* is the compound attenuation factor, and *V* is the compound variance factor. For the predictor unreliability distribution, *MA* is the mean, *VA* is the variance, and *SCVA* is the squared coefficient of variation; variables for criterion unreliability and range restriction are coded similarly. Periods must be inserted for any missing artifact values. The *VARDEF* = *N* option results in the sum of the squared deviations being divided by the number of artifact values, which is what Hunter and Schmidt (1990a) did in their illustrations, rather than the number of artifact values minus one. *VAR_R* is the sample-size-weighted variance calculated with the SAS statements in Table 3.1, and *VAR_E* is the variance attributable to sampling error. Additionally, *RHO* is the population correlation, *VAR_AV* is the variance in the observed coefficients that is attributable to study-to-study variations in the levels of the attenuating artifacts, *VAR_RES1* is the residual observed variance after removal of variance attributable to sampling error only, and *VAR_RES2* is the residual observed variance after removal of variance attributable to sampling error and study-to-study variations in attenuating artifacts. *VAR_RHO* is the variance of the fully corrected correlations in the population, and *PVA_ALL* is the percent of observed variance accounted for by sampling error and study-to-study variations in attenuating artifacts. *PVA_SE* is the percent of variance accounted for by sampling error. *SE_R1* and *SE_R2* are the standard error of the mean correlation for homogeneous and heterogeneous studies respectively. The choice of which of these two is most appropriate must be made by the user for their specific analyses. *L95_CONF* and *U95_CONF* are the lower and upper bounds of the 95% confidence interval. The 1.96 should be replaced with the appropriate *z*-value for the specified interval (i.e., 99% = 2.58; 90% = 1.64). *L95_CRED* and *U95_CRED* are the lower and upper bounds of the 95% credibility interval. Again, the 1.96 should be replaced with the appropriate *z*-value for the specified interval (i.e., 99% = 2.58; 90% = 1.64). *CHI_SQ* is the calculated chi-square test statistic ($df = K$-1) for the homogeneity of the correlations (i.e., is the unexplained variance significantly greater than zero?). The user must refer to a chi-square table to determine the significance of this test at the specified *p* value. A chi-square table is presented in Appendix D of this volume. *CHI_SQ_S* is the calculated chi-square test statistic ($df = k$-1) for correcting for sampling error only (i.e., after computing the sample-weighted correlation.

SAS PRINTOUTS FOR MENTAL ROTATION TEST VALIDITY EXAMPLE

The SAS printout for the mental rotation ability–flight simulator performance example used in Tables 3.1, 3.2, and 3.3 is presented in Table 3.5. This printout includes both the SAS program code and the output (results). The SAS printout for the moderator analysis code shown in Table 3.4 is also shown in Table 3.6.

TABLE 3.5
SAS Printout for the Meta-Analysis of Correlations for Hypothetical Mental
Rotation Test Validity Data

```
1                        The SAS System          01:44 Sunday, April 9, 2000

NOTE: Copyright (c) 1989-1996 by SAS Institute Inc., Cary, NC, USA.
NOTE: SAS (r) Proprietary Software Release 6.09  TS470

**********************************************************************
*                                                                    *
*        Welcome to the new SAS System, Release 6.09 Enhanced        *
*                                                                    *
**********************************************************************

NOTE: SAS system options specified are:
      SORT=4

NOTE: The initialization phase used 0.15 CPU seconds and 2041K.
1           OPTIONS LS=90;
2           DATA OBSERVED;            ***       COMMENTS        **;
3            INPUT STUDY R  N  SEX $;  *** STUDY=STUDY NUMBER, R=CORRELATION **;
4            CARDS;

NOTE: The data set WORK.OBSERVED has 10 observations and 4 variables.
NOTE: The DATA statement used 0.05 CPU seconds and 2679K.

4                             *** N = CORRESPONDING SAMPLE SIZE    **;
15          ;
16          DATA SMP_WGT1;
17           SET OBSERVED;

NOTE: The data set WORK.SMP_WGT1 has 10 observations and 4 variables.
NOTE: The DATA statement used 0.01 CPU seconds and 2683K.

18          PROC MEANS MEAN VAR STD MIN MAX RANGE VARDEF=WEIGHT;
19           VAR R;
20           WEIGHT N;
21           OUTPUT OUT=ADATA MEAN=SWMC  VAR=VAR_R;****************************;

NOTE: The data set WORK.ADATA has 1 observations and 4 variables.
NOTE: The PROCEDURE MEANS printed page 1.
NOTE: The PROCEDURE MEANS used 0.03 CPU seconds and 2888K.

22          DATA SMP_WGT2;                     *** SWMC=SAMPLE-WEIGHTED MEAN R;
23           SET OBSERVED;                     *** MEAN_N=AVERAGE N ACROSS  **;

NOTE: The data set WORK.SMP_WGT2 has 10 observations and 4 variables.
NOTE: The DATA statement used 0.01 CPU seconds and 2888K.

24          PROC MEANS MEAN SUM N MIN MAX RANGE;  *** ALL STUDIES, N=TOTAL N  **;

2                        The SAS System          01:44 Sunday, April 9, 2000

25           VAR N;                        ****K=NUMBER OF STUDIES      **;
26           OUTPUT OUT=BDATA  MEAN=MEAN_N SUM=N  N=K;  ************************;
27

NOTE: The data set WORK.BDATA has 1 observations and 5 variables.
NOTE: The PROCEDURE MEANS printed page 2.
NOTE: The PROCEDURE MEANS used 0.01 CPU seconds and 2888K.

28          DATA ARTIFACT;
29           INPUT RXX RYY U;
30           SWMC=.3011921;
31            A = SQRT (RXX);
32            B = SQRT (RYY);
33            C = SQRT ((1 - (U ** 2)) * (SWMC ** 2) + (U ** 2));
34           CARDS;
```

(continued)

```
NOTE: Missing values were generated as a result of performing an operation on missing
      values.
      Each place is given by: (Number of times) at (Line):(Column).
      2 at 32:10    1 at 33:10    1 at 33:19    1 at 33:24    1 at 33:31    1 at 33:45
      1 at 33:50
NOTE: The data set WORK.ARTIFACT has 5 observations and 7 variables.
NOTE: The DATA statement used 0.03 CPU seconds and 2941K.

40           ;
41           PROC MEANS MEAN VAR N MIN MAX RANGE VARDEF=N;
42              VAR A B C  RXX RYY U;
43              OUTPUT OUT=AFDATA  MEAN=MA MB MC  VAR=VA VB VC;
44

NOTE: The data set WORK.AFDATA has 1 observations and 8 variables.
NOTE: The PROCEDURE MEANS printed page 3.
NOTE: The PROCEDURE MEANS used 0.01 CPU seconds and 2941K.

45           DATA COMPUTE;
46             MERGE ADATA BDATA AFDATA;
47              SCVA = VA / (MA * MA);
48              SCVB = VB / (MB * MB);
49              SCVC = VC / (MC * MC);
50              AA = MA * MB * MC;
51              V = SCVA + SCVB + SCVC;

NOTE: The data set WORK.COMPUTE has 1 observations and 18 variables.
NOTE: The DATA statement used 0.02 CPU seconds and 3029K.

52           PROC PRINT;
53              VAR SCVA SCVB SCVC AA V;
54

NOTE: The PROCEDURE PRINT printed page 4.
NOTE: The PROCEDURE PRINT used 0.01 CPU seconds and 3112K.

55           DATA FINALE;
56             SET COMPUTE;
57              VAR_E = ((1 - (SWMC ** 2)) ** 2) / (MEAN_N - 1);
58              RHO = SWMC / AA;
59              VAR_AV = (RHO ** 2) * (AA ** 2) * V;
60              VAR_RES1 = VAR_R - VAR_E;

3                                     The SAS System           01:44 Sunday, April 9, 2000

61              VAR_RES2 = VAR_R - VAR_E - VAR_AV;
62              VAR_RHO = VAR_RES2 / (AA ** 2);
63           IF VAR_RHO LT 0 THEN VAR_RHO = 0;
64              SD_RHO = SQRT (VAR_RHO);
65              PVA_SE = 100 * (VAR_E / VAR_R);
66              PVA_ALL = 100 * (VAR_AV + VAR_E) / VAR_R;
67              SE_R1 = (1 - (SWMC ** 2)) / SQRT (N - K);
68              SE_R2 = SQRT ((((1 - (SWMC**2)) ** 2) / (N - K)) + (VAR_RES1/K));
69              L95_CONF = SWMC - (1.96 * SE_R1);
70              U95_CONF = SWMC + (1.96 * SE_R1);
71              L95_CRED = RHO - (1.96 * SD_RHO);
72              U95_CRED = RHO + (1.96 * SD_RHO);
73              CHI_SQ_S = (N / ((1 - (SWMC ** 2)) ** 2)) * VAR_R;
74              CHI_SQ = (K * VAR_R) / (VAR_AV + VAR_E);

NOTE: The data set WORK.FINALE has 1 observations and 35 variables.
NOTE: The DATA statement used 0.04 CPU seconds and 3136K.

75           PROC PRINT;
76              VAR   RHO  VAR_RHO  SD_RHO  VAR_AV  VAR_E  VAR_RES1  VAR_RES2  PVA_SE
77                    PVA_ALL  SE_R1  SE_R2  L95_CONF  U95_CONF  L95_CRED  U95_CRED
78                    CHI_SQ_S  CHI_SQ;
NOTE: The PROCEDURE PRINT printed page 5.
NOTE: The PROCEDURE PRINT used 0.01 CPU seconds and 3136K.
```

(continued)

```
NOTE: The SAS session used 0.40 CPU seconds and 3136K.
NOTE: SAS Institute Inc., SAS Campus Drive, Cary, NC USA 27513-2414
                                    The SAS System        01:44 Sunday, April 9, 2000    1
     Analysis Variable : R

           Mean        Variance        Std Dev        Minimum        Maximum         Range
      -------------------------------------------------------------------------------------
        0.3011921      0.0287443       0.1695414      0.0200000      0.6000000      0.5800000
      -------------------------------------------------------------------------------------
                                    The SAS System        01:44 Sunday, April 9, 2000    2
       Analysis Variable : N

                 Mean          Sum    N        Minimum        Maximum          Range
         ---------------------------------------------------------------------------------
          75.5000000    755.0000000   10     25.0000000    125.0000000    100.0000000
         ---------------------------------------------------------------------------------
                                    The SAS System        01:44 Sunday, April 9, 2000    3
         Variable       Mean        Variance  N        Minimum        Maximum          Range
         ---------------------------------------------------------------------------------
         A          0.9028784    0.000810548  5       0.8602325      0.9380832      0.0778506
         B          0.7331663    0.0024672    3       0.6633250      0.7745967      0.1112717
         C          0.7635024    0.0052496    4       0.6635098      0.8557494      0.1922396
         RXX        0.8160000    0.0026240    5       0.7400000      0.8800000      0.1400000
         RYY        0.5400000    0.0050667    3       0.4400000      0.6000000      0.1600000
         U          0.7350000    0.0068750    4       0.6200000      0.8400000      0.2200000
         ---------------------------------------------------------------------------------
                                    The SAS System        01:44 Sunday, April 9, 2000    4
             OBS       SCVA        SCVB         SCVC         AA          V
              1      .00099431   .0045898     .0090054    0.50541    0.014590
                                    The SAS System        01:44 Sunday, April 9, 2000    5
OBS    RHO      VAR_RHO   SD_RHO    VAR_AV     VAR_E    VAR_RES1  VAR_RES2   PVA_SE   PVA_ALL

 1   0.59594   0.063901  0.25279  .0013235  0.011098  0.017646  0.016323  38.6092   43.2136

OBS    SE_R1       SE_R2     L95_CONF   U95_CONF   L95_CRED   U95_CRED   CHI_SQ_S    CHI_SQ

 1   0.033314    0.053614   0.23590    0.36649    0.10047    1.09140    26.2482    23.1408
```

META-ANALYSIS RESULTS SUMMARY TABLE AND INTERPRETATION OF RESULTS FOR MENTAL ROTATION TEST VALIDITY EXAMPLE IN TABLES 3.1 THROUGH 3.6

This section presents a Table 3.7, which summarizes the meta-analysis results for the mental rotation test validity example used in Tables 3.1 through 3.6. So as not to confuse the reader, in the computation of the confidence intervals we used the standard error of the mean correlation for *homogenous* studies (see Equations 3.6 and 3.22) although the results of the meta-analysis—specifically, less than 75% of variance accounted for, a relatively large credibility interval (that does not include zero), and significant chi-square test for homogeneity—clearly suggest that the correlations are not homogeneous and may represent different subpopulations.

TABLE 3.6
SAS Printout for the Meta-Analysis of Correlations—Moderator Analysis for Hypothetical Mental Rotation Test Validity Data With Test Taker Sex (Female) as a Moderator Variable

```
1                               The SAS System        12:42 Saturday, May 20, 2000

NOTE: Copyright (c) 1989-1996 by SAS Institute Inc., Cary, NC, USA.
NOTE: SAS (r) Proprietary Software Release 6.09  TS470
      Licensed to TEXAS A&M UNIVERSITY, Site 0001452001.

*******************************************************************
*                                                                 *
*        Welcome to the new SAS System, Release 6.09 Enhanced     *
*                                                                 *
*******************************************************************

NOTE: SAS system options specified are:
      SORT=4

NOTE: The initialization phase used 0.15 CPU seconds and 2041K.
1            OPTIONS LS=90;
2            DATA OBSERVED;                ***        COMMENTS            **;
3              INPUT STUDY R  N  SEX $;    *** STUDY=STUDY NUMBER, R=CORRELATION **;
4              CARDS;

NOTE: The data set WORK.OBSERVED has 10 observations and 4 variables.
NOTE: The DATA statement used 0.05 CPU seconds and 2679K.

4                               *** N = CORRESPONDING SAMPLE SIZE     **;
15           ;                    ********************************************
16           DATA SMP_WGT1;       ** FOR MALES, CHANGE 'FEMALE' TO 'MALE' HERE **;
17             SET OBSERVED;      **  AND ALSO IN DATA SMP_WGT2 AND RERUN       **;
18             IF SEX='FEMALE';   ********************************************;

NOTE: The data set WORK.SMP_WGT1 has 5 observations and 4 variables.
NOTE: The DATA statement used 0.01 CPU seconds and 2683K.

19           PROC MEANS MEAN VAR STD MIN MAX RANGE VARDEF=WEIGHT;
20             VAR R;
21             WEIGHT N;
22             OUTPUT OUT=ADATA MEAN=SWMC  VAR=VAR_R;

NOTE: The data set WORK.ADATA has 1 observations and 4 variables.
NOTE: The PROCEDURE MEANS printed page 1.
NOTE: The PROCEDURE MEANS used 0.03 CPU seconds and 2888K.

23           DATA SMP_WGT2;                 *******************************;
24             SET OBSERVED;               ** FOR MALES, CHANGE        **;
25             IF SEX='FEMALE';            ** TO 'MALE'                 **;

NOTE: The data set WORK.SMP_WGT2 has 5 observations and 4 variables.
NOTE: The DATA statement used 0.01 CPU seconds and 2888K.

2                               The SAS System        12:42 Saturday, May 20, 2000

26           PROC MEANS MEAN SUM N MIN MAX RANGE;  *******************************;
27             VAR N;                         ** SWMC=SAMPLE-WEIGHTED MEAN R ;
28             OUTPUT OUT=BDATA  MEAN=MEAN_N SUM=N  N=K; * MEAN_N=AVERAGE N ACROSS ;
29                                              ** ALL STUDIES, N=TOTAL N *;

NOTE: The data set WORK.BDATA has 1 observations and 5 variables.
NOTE: The PROCEDURE MEANS printed page 2.
NOTE: The PROCEDURE MEANS used 0.01 CPU seconds and 2888K.

30           DATA ARTIFACT;                       ** K=NUMBER OF STUDIES    *;
31             INPUT RXX RYY U;                   **************************;
32             SWMC=.4219512;
33               A = SQRT (RXX);
34               B = SQRT (RYY);
35               C = SQRT ((1 - (U ** 2)) * (SWMC ** 2) + (U ** 2));
36             CARDS;
```

(continued)

```
NOTE: Missing values were generated as a result of performing an operation on missing
      values.
      Each place is given by: (Number of times) at (Line):(Column).
      2 at 34:10   1 at 35:10   1 at 35:19   1 at 35:24   1 at 35:31   1 at 35:45
      1 at 35:50
NOTE: The data set WORK.ARTIFACT has 5 observations and 7 variables.
NOTE: The DATA statement used 0.03 CPU seconds and 2941K.

42          ;
43             PROC MEANS MEAN VAR N MIN MAX RANGE VARDEF=N;
44                VAR A B C  RXX RYY U;
45                OUTPUT OUT=AFDATA  MEAN=MA MB MC  VAR=VA VB VC;
46

NOTE: The data set WORK.AFDATA has 1 observations and 8 variables.
NOTE: The PROCEDURE MEANS printed page 3.
NOTE: The PROCEDURE MEANS used 0.01 CPU seconds and 2941K.

47             DATA COMPUTE;
48                MERGE ADATA BDATA AFDATA;
49                   SCVA = VA / (MA * MA);
50                   SCVB = VB / (MB * MB);
51                   SCVC = VC / (MC * MC);
52                   AA = MA * MB * MC;
53                   V = SCVA + SCVB + SCVC;

NOTE: The data set WORK.COMPUTE has 1 observations and 18 variables.
NOTE: The DATA statement used 0.02 CPU seconds and 3029K.

54             PROC PRINT;
55                VAR SCVA SCVB SCVC AA V;
56

NOTE: The PROCEDURE PRINT printed page 4.
NOTE: The PROCEDURE PRINT used 0.01 CPU seconds and 3112K.

57             DATA FINALE;
58                SET COMPUTE;
59                   VAR_E = ((1 - (SWMC ** 2)) ** 2) / (MEAN_N - 1);
60                   RHO = SWMC / AA;
3                                        The SAS System         12:42 Saturday, May 20, 2000

61                   VAR_AV = (RHO ** 2) * (AA ** 2) * V;
62                   VAR_RES1 = VAR_R - VAR_E;
63                   VAR_RES2 = VAR_R - VAR_E - VAR_AV;
64                   VAR_RHO = VAR_RES2 / (AA ** 2);
65               IF VAR_RHO LT 0 THEN VAR_RHO = 0;
66                   SD_RHO = SQRT (VAR_RHO);
67                   PVA_SE = 100 * (VAR_E / VAR_R);
68                   PVA_ALL = 100 * (VAR_AV + VAR_E) / VAR_R;
69                   SE_R1 = (1 - (SWMC ** 2)) / SQRT (N - K);
70                   SE_R2 = SQRT ((((1 - (SWMC**2)) ** 2) / (N - K)) + (VAR_RES1/K));
71                   L95_CONF = SWMC - (1.96 * SE_R1);
72                   U95_CONF = SWMC + (1.96 * SE_R1);
73                   L95_CRED = RHO - (1.96 * SD_RHO);
74                   U95_CRED = RHO + (1.96 * SD_RHO);
75                   CHI_SQ_S = (N / ((1 - (SWMC ** 2)) ** 2)) * VAR_R;
76                   CHI_SQ = (K * VAR_R) / (VAR_AV + VAR_E);

NOTE: The data set WORK.FINALE has 1 observations and 35 variables.
NOTE: The DATA statement used 0.04 CPU seconds and 3136K.

77             PROC PRINT;
78                VAR   RHO  VAR_RHO  SD_RHO  VAR_AV  VAR_E  VAR_RES1  VAR_RES2  PVA_SE
79                      PVA_ALL  SE_R1  SE_R2  L95_CONF  U95_CONF  L95_CRED  U95_CRED
80                      CHI_SQ_S  CHI_SQ;
NOTE: The PROCEDURE PRINT printed page 5.
NOTE: The PROCEDURE PRINT used 0.01 CPU seconds and 3136K.
```

(continued)

```
NOTE: The SAS session used 0.40 CPU seconds and 3136K.
NOTE: SAS Institute Inc., SAS Campus Drive, Cary, NC USA 27513-2414
                              The SAS System     12:42 Saturday, May 20, 2000    1

    Analysis Variable : R


           Mean        Variance      Std Dev      Minimum       Maximum        Range
     ---------------------------------------------------------------------------------
         0.4219512    0.0166157    0.1289019    0.2800000    0.6000000    0.3200000
     ---------------------------------------------------------------------------------
                              The SAS System     12:42 Saturday, May 20, 2000    2

       Analysis Variable : N


              Mean         Sum   N      Minimum       Maximum        Range
        --------------------------------------------------------------------------
         82.0000000   410.0000000   5   25.0000000   125.0000000   100.0000000
        --------------------------------------------------------------------------
                              The SAS System     12:42 Saturday, May 20, 2000    3

     Variable          Mean       Variance   N      Minimum       Maximum        Range
     -------------------------------------------------------------------------------------
       A            0.9028784   0.000810548   5    0.8602325    0.9380832    0.0778506
       B            0.7331663   0.0024672     3    0.6633250    0.7745967    0.1112717
       C            0.7897699   0.0039991     4    0.7028536    0.8706410    0.1677874
       RXX          0.8160000   0.0026240     5    0.7400000    0.8800000    0.1400000
       RYY          0.5400000   0.0050667     3    0.4400000    0.6000000    0.1600000
       U            0.7350000   0.0068750     4    0.6200000    0.8400000    0.2200000
     -------------------------------------------------------------------------------------
                              The SAS System     12:42 Saturday, May 20, 2000    4

              OBS        SCVA         SCVB         SCVC          AA          V

               1       .00099431    .0045898     .0064116     0.52280     0.011996
                              The SAS System     12:42 Saturday, May 20, 2000    5

  OBS    RHO      VAR_RHO    SD_RHO    VAR_AV     VAR_E    VAR_RES1   VAR_RES2    PVA_SE    PVA_ALL

   1   0.80710   0.022461   0.14987   .0021357   .0083409  .0082748   .0061391   50.1989   63.0527

  OBS    SE_R1       SE_R2     L95_CONF   U95_CONF   L95_CRED   U95_CRED   CHI_SQ_S    CHI_SQ

   1   0.040843   0.057647   0.34190    0.50200    0.51336    1.10085    10.0833     7.92988
```

Note. This printout is for females only. The program will have to be rerun for males by changing the level of the moderator variable in the *IF* statements. (The SAS code for all the moderators used in this example can be found at www.erlbaum.com.) An alternative is to run the analyses for both levels of the moderator in the same program by repeating or cutting and pasting the males code to the end of the females code. Running all levels of the moderator in the same program can also be accomplished by using *PROC SORT* or *WHERE* statements.

Under these circumstances the confidence interval should be computed using the standard error of the mean correlation for *heterogenous* studies (see Equations 3.8 and 3.23). However, the standard error for homogenous correlations was used because this is what was used in the examples in preceding sections.

It is important to note that, on occasion, sampling error and artifactual variance will be larger than the observed variance across the effect sizes from the studies, causing more than 100% of the observed variance to be accounted for by

TABLE 3.7
Meta-Analysis Results for Mental Rotation Ability–Flight Simulator Performance Example

VAR.	# of Data Pnts. (k)	Total Sample Size	Sample-weighted Mean r	Corr. Mean r (ρ)	Corr. SD (SDρ)	% Var. due to Artifacts	95% Conf. Interval	95% Cred. Interval	χ²
Overall	10	755	0.30	0.60	0.25	43.21	0.24 : 0.37	0.10 : 1.00	23.14*
Sex									
Female	5	410	0.42	0.80	0.15	63.05	0.34 : 0.50	0.51: 1.00	7.93
Male	5	345	0.16	0.30	0.00	100.00	0.05 : 0.26	0.32 : 0.32	1.82

Note. "VAR." = variable; "Pnts." = points; "Corr." = corrected; "Var." = variance; "Conf."
= confidence; "Cred." = credibility. *$p<.05$, $df = k$-1.

artifacts. A key point is that the data set for a particular meta-analysis represents
only one possible set of studies taken from the target population. As such, the
variability across these studies does not always exactly represent the true variability in the population. When the variability underestimates the true population
variability, the possibility exists that the estimate of artifactual variance will exceed the observed variance. When this happens, the logical course of action (and
standard practice) is to substitute 100% for the actual percent-of-variance estimate. Doing so provides for a clearer result and does not change the basic premise that no moderator variables are operating.

Interpretation of Results Presented in Table 3.7

Results of the meta-analysis presented in Table 3.7 show that the overall sample-
weighted mean correlation was 0.30. Correcting for predictor and criterion reliability and range restriction resulted in a corrected mean *r* of 0.60. However, because the corrected variance was reasonably large, coupled with a significant chi-
square test of homogeneity and a large credibility interval with artifacts account-
ing for only 43% of the variance, the decision was made to test for test taker sex
as a moderator variable.

The results of the moderator analysis, which are also presented in Table 3.7,
clearly indicate that test taker sex moderates that mental rotation test–flight simulator performance relationship. For both males and females, substantially more

TABLE 3.8
Artifact Distribution for Mental Rotation Test Example

Variable	k	Mean	Variance
r_{xx}	5	0.82	0.0026
r_{yy}	3	0.54	0.0051
U	4	0.74	0.0069

Note. r_{xx} = predictor reliability; r_{yy} = criterion reliability; U = range restriction; k = number of data points contributing to artifact distribution; *Mean* = average of artifact distribution; *Variance* = variance of the artifact distribution.

artifactual variance was accounted for, the chi-square tests of homogeneity were not significant, and the width of the 95% credibility intervals were smaller than those for the overall data. Also, credibility intervals did not include zero. The corrected variance was also smaller for both females and males, compared to the overall. Finally, the validity of the test was substantially higher for females. In conclusion, these results indicate that the mental rotation test is generally a valid predictor of simulator flight performance, but that this validity appears to be a function of the sex of the test taker, with the test displaying somewhat higher validities for female test takers than for male test takers.

Finally, summary information about the artifact distribution used in the mental rotation test example is shown in Table 3.8. Next, Table 3.9 shows the squared coefficient of variation, compound attenuation factor, and compound variance factor values.

Fully Hierarchical Moderator Analysis

Fully hierarchical moderator analysis is, basically, an investigation of interactions among the moderators. For instance, we might be interested in investigating whether mental rotation test validation studies that used male pilots resulted in different predictor–criterion relationships than those that used female nonpilots. Fully hierarchical moderator analyses are technically superior to partial or incomplete moderator analyses. However, the most frequent, albeit not the best

TABLE 3.9
Squared Coefficient of Variation, Compound Attenuation Factor,
and Compound Variance Factor Values for Mental Rotation Test
Example

FACTOR	Overall	Females	Males
SCV(A)	0.00099431	0.00099431	0.00099431
SCV(B)	0.0045898	0.0045898	0.0045898
SCV(C)	0.0090054	0.0064116	0.011580
AA	0.50541	0.52280	0.49180
V	0.014590	0.011996	0.017164

Note. $SCV(A)$ = squared coefficient of variation for predictor reliability; $SCV(B)$ = squared coefficient of variation for criterion reliability; $SCV(C)$ = squared coefficient of variation for range restriction; AA = compound attenuation factor; V = compound variance factor.

practice in testing for moderators, is to first include all the studies in an overall meta-analysis and then break the studies down by one moderator (e.g., test taker sex—see Table 3.7), and then by another moderator (e.g., pilot vs. nonpilot), and so on. This approach to moderator analysis is not fully hierarchical because the moderators are not considered in combination, but instead, they are considered independently of one another. Failure to consider moderator variables in combination can result in major errors of interpretation because one fails to take into account interaction and confounding effects (Hunter & Schmidt, 1990a).

The implementation of a fully hierarchical moderator analysis first calls for the creation of a matrix that represents all levels of all moderators. The data points are then distributed across the cells of the matrix. Thus, the feasibility of fully hierarchical moderator analyses is primarily a function of the number of data points in each cell and is constrained by the total number of data points, the number of moderators, and the number of levels of the moderators. Although technically superior, fully hierarchical moderator analyses are not very common in the meta-analysis literature because typically, when the meta-analytic data are disaggregated to this level, the number of data points in each cell is usually quite small and the meta-analysis of such small numbers of data points is very questionable as it raises major, serious concerns about the stability and interpretability of meta-analytic estimates. For instance, in Table 3.7, disaggregating the data to just two levels of a single moderator resulted in five data points per cell. And in meta-analytic terms, five (and even ten) data points is a relatively small sample of data points.

Specifically, although this goal is rarely met, a meta-analysis ideally calls for several hundred data points. This issue is further exacerbated by moderator, and even more so, by fully hierarchical moderator analysis because disaggregating variables down to sublevels sometimes drastically reduces the number of correlations to be meta-analyzed to a relatively small number. And as the number of studies in a meta-analysis decreases, the likelihood of sampling error increases. Consequently, when one corrects for sampling error, it becomes more likely to obtain favorable than unfavorable results. A meta-analysis of a small number of studies are biased in favor of obtaining "positive" results. Thus, they are to be discouraged, and if unavoidable, should be cautiously interpreted because they could be very misleading. Although there is no magic cutoff as to the minimum number of studies, conceptually, it is difficult to make a case that a meta-analysis of five or ten data points represents a population parameter.

The SAS code required to run a fully hierarchical moderator analysis is quite simple. Specifically, we can run a fully hierarchical moderator analysis by simultaneously analyzing additional moderator variables by using joint or compound *IF* statements (e.g., *IF SEX* = *'FEMALE' AND STATUS* = *'PILOT';*). SAS code necessary to run a fully hierarchical moderator analysis on another hypothetical mental rotation test data set is presented in Table 3.10. The two moderators of the test taker sex (male or female) and status (pilot vs. nonpilot). The SAS printout for Table 3.10 is presented in Table 3.11, and a summary of the meta-analysis results is presented in Table 3.12.

Note that for the code presented in Table 3.10, the mean sample-weighted correlation—*SWMC*—obtained from program code lines 129-132 must be inserted in line 142. Thus, the code presented in Table 3.10 must be run first, and then the *SWMC* value inserted in line 142 and the program rerun.

TABLE 3.10

SAS Statements for Conducting a Fully Hierarchical Moderator Analysis for Hypothetical Mental Rotation Test Validity Data With Test Taker Sex (Female) and Status (Pilot) as Moderator Variables for the Meta-Analysis of Correlations

001	DATA OBSERVED;				
002	INPUT STUDY R N SEX $ STATUS $;				
003					
004	CARDS;				
005	1	0.60	85	FEMALE	PILOT
006	2	0.12	70	MALE	NONPILOT
007	3	0.24	125	MALE	PILOT
008	4	0.02	25	MALE	PILOT
009	5	0.50	75	FEMALE	NONPILOT
010	6	0.35	100	FEMALE	NONPILOT
011	7	0.16	75	MALE	NONPILOT
012	8	0.28	125	FEMALE	PILOT
013	9	0.07	50	MALE	PILOT
014	10	0.58	25	FEMALE	NONPILOT
015	11	0.12	97	MALE	NONPILOT
016	12	0.60	9	MALE	PILOT
017	13	0.25	69	MALE	PILOT
018	14	0.17	68	MALE	NONPILOT
019	15	0.57	64	FEMALE	NONPILOT
020	16	0.60	9	MALE	PILOT
021	17	0.34	64	FEMALE	NONPILOT
022	18	0.33	78	FEMALE	NONPILOT
023	19	0.04	87	MALE	PILOT
024	20	0.05	35	MALE	PILOT
025	21	0.51	60	FEMALE	NONPILOT
026	22	0.33	46	FEMALE	NONPILOT
027	23	0.53	70	FEMALE	NONPILOT
028	24	0.05	64	FEMALE	PILOT
029	25	0.59	24	MALE	NONPILOT
030	26	0.19	83	MALE	NONPILOT
031	27	0.18	31	MALE	NONPILOT
032	28	0.43	54	FEMALE	PILOT
033	29	0.60	68	FEMALE	PILOT
034	30	0.15	18	MALE	NONPILOT
035	31	0.43	21	FEMALE	PILOT
036	32	0.26	19	MALE	PILOT
037	33	0.35	34	FEMALE	NONPILOT
038	34	0.19	24	MALE	NONPILOT
039	35	0.30	52	FEMALE	NONPILOT
040	36	0.52	40	FEMALE	NONPILOT
041	37	0.40	87	FEMALE	PILOT
042	38	0.37	29	MALE	NONPILOT
043	39	0.36	35	MALE	NONPILOT
044	40	0.24	42	MALE	PILOT
045	41	0.45	44	FEMALE	PILOT
046	42	0.32	71	FEMALE	NONPILOT
047	43	0.58	75	FEMALE	NONPILOT
048	44	0.58	64	FEMALE	NONPILOT
049	45	0.37	40	FEMALE	NONPILOT
050	46	0.19	34	MALE	NONPILOT

(continued)

For the fully hierarchical moderator analysis, we used the same artifact distribution as used in the example presented in Tables 3.1-3.8. A summary of this information is presented in Tables 3.13 and 3.14.

051	47	0.25	19	FEMALE	PILOT
052	48	0.30	15	MALE	NONPILOT
053	49	0.42	51	FEMALE	PILOT
054	50	0.11	37	MALE	NONPILOT
055	51	0.11	35	MALE	NONPILOT
056	52	0.54	85	FEMALE	NONPILOT
057	53	0.19	9	MALE	NONPILOT
058	54	0.58	37	FEMALE	NONPILOT
059	55	0.56	65	FEMALE	NONPILOT
060	56	0.36	68	FEMALE	NONPILOT
061	57	0.04	28	MALE	PILOT
062	58	0.09	27	MALE	PILOT
063	59	0.32	47	FEMALE	NONPILOT
064	60	0.27	44	FEMALE	PILOT
065	61	0.12	62	MALE	NONPILOT
066	62	0.41	35	FEMALE	PILOT
067	63	0.26	63	FEMALE	PILOT
068	64	0.09	25	MALE	PILOT
069	65	0.29	73	FEMALE	PILOT
070	66	0.13	6	MALE	NONPILOT
071	67	0.36	41	FEMALE	NONPILOT
072	68	0.46	8	FEMALE	PILOT
073	69	0.28	37	FEMALE	PILOT
074	70	0.04	26	MALE	PILOT
075	71	0.33	59	FEMALE	NONPILOT
076	72	0.22	36	MALE	PILOT
077	73	0.02	83	MALE	PILOT
078	74	0.44	89	FEMALE	PILOT
079	75	0.58	16	FEMALE	NONPILOT
080	76	0.28	48	MALE	PILOT
081	77	0.59	7	FEMALE	NONPILOT
082	78	0.44	102	FEMALE	PILOT
083	79	0.07	75	MALE	PILOT
084	80	0.12	49	MALE	NONPILOT
085	81	0.17	46	MALE	NONPILOT
086	82	0.38	52	FEMALE	NONPILOT
087	83	0.27	55	MALE	PILOT
088	84	0.05	62	MALE	PILOT
089	85	0.23	28	MALE	PILOT
090	86	0.44	63	MALE	PILOT
091	87	0.13	58	MALE	NONPILOT
092	88	0.10	21	MALE	NONPILOT
093	89	0.35	40	FEMALE	NONPILOT
094	90	0.17	22	MALE	NONPILOT
095	91	0.21	35	FEMALE	PILOT
096	92	0.36	35	FEMALE	NONPILOT
097	93	0.04	9	MALE	PILOT
098	94	0.28	66	FEMALE	PILOT
099	95	0.57	22	FEMALE	NONPILOT
100	96	0.33	63	FEMALE	NONPILOT
111	97	0.46	59	MALE	PILOT
112	98	0.56	55	FEMALE	NONPILOT
113	99	0.36	75	FEMALE	NONPILOT
114	100	0.12	20	MALE	NONPILOT
115	101	0.21	12	MALE	PILOT
116	102	0.44	40	FEMALE	PILOT
117	103	0.08	56	MALE	PILOT
118	104	0.12	79	MALE	NONPILOT
119	105	0.18	50	MALE	NONPILOT
120	106	0.41	31	FEMALE	PILOT
121	107	0.27	32	FEMALE	PILOT
122	108	0.03	46	MALE	NONPILOT
123	109	0.18	49	FEMALE	NONPILOT
124	110	0.27	5	FEMALE	PILOT
125	;				

(continued)

```
126   DATA SMP_WGT1;
127     SET OBSERVED;
128   IF SEX='FEMALE' AND STATUS='PILOT';
129   PROC MEANS MEAN VAR STD MIN MAX RANGE VARDEF=WEIGHT;
130     VAR R;
131     WEIGHT N;
132     OUTPUT OUT=ADATA MEAN=SWMC VAR=VAR_R;
133   DATA SMP_WGT2;
134     SET OBSERVED;
135   IF SEX='FEMALE' AND STATUS='PILOT';
136   PROC MEANS MEAN SUM N MIN MAX RANGE;
137     VAR N;
138     OUTPUT OUT=BDATA MEAN=MEAN_N SUM=N N=K;
139
140   DATA ARTIFACT;
141     INPUT RXX RYY U;
142     SWMC=.3685524;
143     A = SQRT (RXX);
144     B = SQRT (RYY);
145     C = SQRT ((1 - (U ** 2)) * (SWMC ** 2) + (U ** 2));
146   CARDS;
147   .88 .60 .78
148   .74 .44 .84
149   .82 .58 .70
150   .86 .  .62
151   .78 .  .
152   ;
153   PROC MEANS MEAN VAR N MIN MAX RANGE VARDEF=N;
154     VAR A B C  RXX RYY U;
155     OUTPUT OUT=AFDATA  MEAN=MA MB MC  VAR=VA VB VC;
156
157   DATA COMPUTE;
158     MERGE ADATA BDATA AFDATA;
159     SCVA = VA / (MA * MA);
160     SCVB = VB / (MB * MB);
161     SCVC = VC / (MC * MC);
162     AA = MA * MB * MC;
163     V = SCVA + SCVB + SCVC;
164   PROC PRINT;
165     VAR SCVA SCVB SCVC AA V;
166
167   DATA FINALE;
168     SET COMPUTE;
169     VAR_E = ((1 - (SWMC ** 2)) ** 2) / (MEAN_N - 1);
170     RHO = SWMC / AA;
171     VAR_AV = (RHO ** 2) * (AA ** 2) * V;
172     VAR_RES1 = VAR_R - VAR_E;
173     VAR_RES2 = VAR_R - VAR_E - VAR_AV;
174     VAR_RHO = VAR_RES2 / (AA ** 2);
175   IF VAR_RHO LT 0 THEN VAR_RHO = 0;
176     SD_RHO = SQRT (VAR_RHO);
177     PVA_SE = 100 * (VAR_E / VAR_R);
178     PVA_ALL = 100 * (VAR_AV + VAR_E) / VAR_R;
179     SE_R1 = (1 - (SWMC ** 2)) / SQRT (N - K);
180     SE_R2 = SQRT (((((1 - (SWMC**2)) ** 2) / (N - K)) + (VAR_RES1/K));
181     L95_CONF = SWMC - (1.96 * SE_R1);
182     U95_CONF = SWMC + (1.96 * SE_R1);
183     L95_CRED = RHO - (1.96 * SD_RHO);
184     U95_CRED = RHO + (1.96 * SD_RHO);
185     CHI_SQ_S = (N / ((1 - (SWMC ** 2)) ** 2)) * VAR_R;
186     CHI_SQ = (K * VAR_R) / (VAR_AV + VAR_E);
```

(continued)

```
187 | PROC PRINT;
188 | VAR RHO VAR_RHO SD_RHO VAR_AV VAR_E VAR_RES1 VAR_RES2 PVA_SE
189 |     PVA_ALL SE_R1 SE_R2 L95_CONF U95_CONF L95_CRED U95_CRED
190 |     CHI_SQ_S CHI_SQ;
```

Note. The mean sample-weighted correlation—*SWMC*—obtained from program code lines 129-132 must be inserted in line 142. Thus, the code presented in Table 3.10 must be run first, and then the *SWMC* value inserted in line 142 and the program rerun. The above SAS code is for females and pilots only. To run the meta-analysis for the other moderators, the joint levels of the moderator variables will have to be selected in the *IF* statements and the program rerun. (The SAS code for all moderators used in this example can be found at www.erlbaum.com.) An alternative is to run the analyses for all combinations of the moderator variables in the same program by repeating or cutting and pasting the code for the other moderator variable levels to the end of the code presented in Table 3.10. Running all levels of the moderators in the same program can also be accomplished by using *PROC SORT* or *WHERE* statements.

Interpretation of Results Presented in Table 3.12

Results of the meta-analysis presented in Table 3.12 show that the overall sample-weighted mean correlation was 0.30. Correcting for predictor and criterion reliability and range restriction resulted in a corrected mean *r* of 0.60. Approximately 62% of the variance due to statistical artifacts was accounted for. The possible operation of moderator variables was suggested by the fairly large 95% credibility interval (0.22-0.98; although it did not include zero), and the significant chi-square test of homogeneity. Thus, although there is a reasonably moderate relationship between performance on the mental rotation test and flight simulator performance, this relationship seems to be influenced by moderators.

This was supported by the results of a subsequent fully hierarchical moderator analysis of test taker sex and status. For the results of these moderator analyses, there was no variability in the corrected correlations, all the variance due statistical artifacts was accounted for, the 95% credibility intervals were small and did not include zero, and the chi-square tests of homogeneity were not significant. The results also showed higher validities for female vs. male test takers. In addition, for females, higher validities were obtained for nonpilots ($\bar{r}$ = 0.82 vs. 0.37). In contrast, for males, higher validities were obtained for pilots ($\bar{r}$ = 0.36 vs. 0.33), although the difference was smaller than that obtained for female pilots vs. nonpilots.

TABLE 3.11

SAS Printout for the Meta-Analysis of Correlations—Fully Hierarchical Moderator Analysis for Hypothetical Mental Rotation Test Validity Data With Test Taker Sex (Female) and Status (Pilot) as the Moderator Variables

```
1                        The SAS System        20:54 Sunday, May 21, 2000

NOTE: Copyright (c) 1989-1996 by SAS Institute Inc., Cary, NC, USA.
NOTE: SAS (r) Proprietary Software Release 6.09  TS470

************************************************************************
*                                                                      *
*         Welcome to the new SAS System, Release 6.09 Enhanced         *
*                                                                      *
************************************************************************

NOTE: SAS system options specified are:
      SORT=4

NOTE: The initialization phase used 0.14 CPU seconds and 2041K.
1           OPTIONS LS=90;
2           DATA OBSERVED;
3             INPUT STUDY R  N  SEX $  STATUS $;
4
5           CARDS;

NOTE: The data set WORK.OBSERVED has 110 observations and 5 variables.
NOTE: The DATA statement used 0.05 CPU seconds and 2679K.

5
116           ;
117         DATA SMP_WGT1;
118           SET OBSERVED;
119           IF SEX='FEMALE' AND STATUS='PILOT';

NOTE: The data set WORK.SMP_WGT1 has 24 observations and 5 variables.
NOTE: The DATA statement used 0.01 CPU seconds and 2683K.

120         PROC MEANS MEAN VAR STD MIN MAX RANGE VARDEF=WEIGHT;
121           VAR R;
122           WEIGHT N;
123           OUTPUT OUT=ADATA MEAN=SWMC  VAR=VAR_R;

NOTE: The data set WORK.ADATA has 1 observations and 4 variables.
NOTE: The PROCEDURE MEANS printed page 1.
NOTE: The PROCEDURE MEANS used 0.03 CPU seconds and 2888K.

124         DATA SMP_WGT2;
125           SET OBSERVED;
126           IF SEX='FEMALE' AND STATUS='PILOT';

NOTE: The data set WORK.SMP_WGT2 has 24 observations and 5 variables.

2                        The SAS System        20:54 Sunday, May 21, 2000

NOTE: The DATA statement used 0.01 CPU seconds and 2888K.

127         PROC MEANS MEAN SUM N MIN MAX RANGE;
128           VAR N;
129           OUTPUT OUT=BDATA  MEAN=MEAN_N SUM=N  N=K;
130

NOTE: The data set WORK.BDATA has 1 observations and 5 variables.
NOTE: The PROCEDURE MEANS printed page 2.
NOTE: The PROCEDURE MEANS used 0.01 CPU seconds and 2888K.

131         DATA ARTIFACT;
132           INPUT RXX RYY U;
133           SWMC=.3685524;
134           A = SQRT (RXX);
135           B = SQRT (RYY);
136           C = SQRT ((1 - (U ** 2)) * (SWMC ** 2) + (U ** 2));
137           CARDS;
```

(continued)

```
NOTE: Missing values were generated as a result of performing an operation on missing
      values.
      Each place is given by: (Number of times) at (Line):(Column).
      2 at 135:10    1 at 136:10    1 at 136:19    1 at 136:24    1 at 136:31    1 at 136:45
      1 at 136:50
NOTE: The data set WORK.ARTIFACT has 5 observations and 7 variables.
NOTE: The DATA statement used 0.03 CPU seconds and 2941K.

143          ;
144          PROC MEANS MEAN VAR N MIN MAX RANGE VARDEF=N;
145             VAR A B C  RXX RYY U;
146             OUTPUT OUT=AFDATA  MEAN=MA MB MC  VAR=VA VB VC;
147

NOTE: The data set WORK.AFDATA has 1 observations and 8 variables.
NOTE: The PROCEDURE MEANS printed page 3.
NOTE: The PROCEDURE MEANS used 0.01 CPU seconds and 2941K.

148          DATA COMPUTE;
149             MERGE ADATA BDATA AFDATA;
150                SCVA = VA / (MA * MA);
151                SCVB = VB / (MB * MB);
152                SCVC = VC / (MC * MC);
153                AA = MA * MB * MC;
154                V = SCVA + SCVB + SCVC;

NOTE: The data set WORK.COMPUTE has 1 observations and 18 variables.
NOTE: The DATA statement used 0.02 CPU seconds and 3029K.

155          PROC PRINT;
156             VAR SCVA SCVB SCVC AA V;
157

NOTE: The PROCEDURE PRINT printed page 4.
NOTE: The PROCEDURE PRINT used 0.01 CPU seconds and 3112K.

158          DATA FINALE;
159             SET COMPUTE;
160                VAR_E = ((1 - (SWMC ** 2)) ** 2) / (MEAN_N - 1);

3                              The SAS System                20:54 Sunday, May 21, 2000

161                RHO = SWMC / AA;
162                VAR_AV = (RHO ** 2) * (AA ** 2) * V;
163                VAR_RES1 = VAR_R - VAR_E;
164                VAR_RES2 = VAR_R - VAR_E - VAR_AV;
165                VAR_RHO = VAR_RES2 / (AA ** 2);
166          IF VAR_RHO LT 0 THEN VAR_RHO = 0;
167                SD_RHO = SQRT (VAR_RHO);
168                PVA_SE = 100 * (VAR_E / VAR_R);
169                PVA_ALL = 100 * (VAR_AV + VAR_E) / VAR_R;
170                SE_R1 = (1 - (SWMC ** 2)) / SQRT (N - K);
171                SE_R2 = SQRT ((((1 - (SWMC**2)) ** 2) / (N - K)) + (VAR_RES1/K));
172                L95_CONF = SWMC - (1.96 * SE_R1);
173                U95_CONF = SWMC + (1.96 * SE_R1);
174                L95_CRED = RHO - (1.96 * SD_RHO);
175                U95_CRED = RHO + (1.96 * SD_RHO);
176                CHI_SQ_S = (N / ((1 - (SWMC ** 2)) ** 2)) * VAR_R;
177                CHI_SQ = (K * VAR_R) / (VAR_AV + VAR_E);

NOTE: The data set WORK.FINALE has 1 observations and 35 variables.
NOTE: The DATA statement used 0.04 CPU seconds and 3136K.

178          PROC PRINT;
179             VAR    RHO   VAR_RHO  SD_RHO  VAR_AV  VAR_E  VAR_RES1  VAR_RES2  PVA_SE
180                    PVA_ALL  SE_R1  SE_R2  L95_CONF  U95_CONF  L95_CRED  U95_CRED
181                    CHI_SQ_S  CHI_SQ;
NOTE: The PROCEDURE PRINT printed page 5.
NOTE: The PROCEDURE PRINT used 0.01 CPU seconds and 3136K.

NOTE: The SAS session used 0.41 CPU seconds and 3136K.
NOTE: SAS Institute Inc., SAS Campus Drive, Cary, NC USA 27513-2414
                               The SAS System                20:54 Sunday, May 21, 2000   1
```

Analysis Variable : R

Mean	Variance	Std Dev	Minimum	Maximum	Range
0.3685524	0.0170268	0.1304867	0.0500000	0.6000000	0.5500000

(continued)

```
                                    The SAS System        20:54 Sunday, May 21, 2000    2

          Analysis Variable : N


                   Mean          Sum    N      Minimum       Maximum          Range
          -----------------------------------------------------------------------------
             53.2500000      1278.00    24    5.0000000   125.0000000   120.0000000
          -----------------------------------------------------------------------------
                                    The SAS System        20:54 Sunday, May 21, 2000    3

          Variable          Mean       Variance   N       Minimum       Maximum          Range
          --------------------------------------------------------------------------------------
          A            0.9028784    0.000810548    5     0.8602325     0.9380832      0.0778506
          B            0.7331663    0.0024672      3     0.6633250     0.7745967      0.1112717
          C            0.7772054    0.0045696      4     0.6841180     0.8634747      0.1793567
          RXX          0.8160000    0.0026240      5     0.7400000     0.8800000      0.1400000
          RYY          0.5400000    0.0050667      3     0.4400000     0.6000000      0.1600000
          U            0.7350000    0.0068750      4     0.6200000     0.8400000      0.2200000
          --------------------------------------------------------------------------------------
                                    The SAS System        20:54 Sunday, May 21, 2000    4

                   OBS        SCVA         SCVB         SCVC        AA           V

                    1      .00099431    .0045898     .0075650    0.51448    0.013149
                                    The SAS System        20:54 Sunday, May 21, 2000    5

    OBS    RHO    VAR_RHO   SD_RHO   VAR_AV   VAR_E   VAR_RES1  VAR_RES2  PVA_SE  PVA_ALL

     1  0.71636  .0035821  0.059850  .0017860  0.014293  .0027342  .00094813  83.9419  94.4315

    OBS    SE_R1       SE_R2      L95_CONF   U95_CONF   L95_CRED   U95_CRED   CHI_SQ_S    CHI_SQ

     1   0.024403    0.026635    0.32072    0.41638    0.59905    0.83367    29.1384    25.4152
```

Note. This printout is for females and pilots only. The program will have to be rerun for the other combinations of the moderator variable (see Table 3.12) by changing the levels of the joint moderator variables in the *IF* statements. (The SAS code for all the moderators used in this example can be found at www.erlbaum.com.) An alternative is to run the analyses for all combinations of the moderator in the same program by repeating or cutting and pasting the code for the other moderator variables to the end of the code in Tables 3.10 and 3.11. Running all combinations of the moderator in the same program can also be accomplished by using *PROC SORT* or *WHERE* statements.

    As previously noted, summary information for the artifact distribution used in the fully hierarchical moderator analysis example is presented in Table 3.13. Because we used the same distribution as before, this information is the same as that presented in Table 3.8. Table 3.14 presents the squared coefficient of variation, the compound attenuation factor, and the compound variance factor values for the fully hierarchical moderator analysis.

SUMMARY

This chapter presented the SAS PROC MEANS procedure in the meta-analysis of correlations, focusing primarily on the Hunter and Schmidt (1990a) approach.

TABLE 3.12
Meta-Analysis Results for Mental Rotation Ability–Flight Simulator
Performance Example

VAR.	# of Data Pnts. (k)	Total Sample Size	Sample-weighted Mean r	Corr. Mean r (ρ)	Corr. SD ($SD\rho$)	% Var. due to Artifacts	95% Conf. Interval	95% Cred. Interval	χ^2
Overall	100	5357	0.30	0.60	0.19	61.38	0.28 : 0.33	0.22 : 0.98	166.39*
Sex									
Female	57	3052	0.40	0.77	0.00	100.00	0.37 : 0.43	0.77 : 0.77	56.69
Male	53	2305	0.18	0.35	0.00	100.00	0.13 : 0.21	0.35 : 0.35	33.79
Status									
Pilot	50	2440	0.28	0.55	0.19	63.62	0.24 : 0.31	0.18 : 0.92	73.81*
Non-pilot	60	2917	0.32	0.64	0.19	67.07	0.29 : 0.36	0.27 : 1.00	89.45*
Sex/ Status									
Female/ Pilot	24	1278	0.37	0.72	0.06	94.43	0.32 : 0.42	0.60 : 0.83	25.42
Female/ Non-pilot	33	1774	0.43	0.82	0.00	100.00	0.39 : 0.47	0.82 : 0.82	28.04
Male/ Pilot	26	1162	0.18	0.36	0.00	100.00	0.12 : 0.23	0.36 : 0.36	24.39
Male/ Non-pilot	27	1143	0.16	0.33	0.00	100.00	0.11 : 0.22	0.33 : 0.33	9.37

Note. "VAR." = variable; "Pnts.." = points; "Corr." = corrected; "Var." = variance; "Conf." = confidence; "Cred." = credibility. *$p<.05$, $df = k-1$.

TABLE 3.13
Artifact Distribution for Fully
Hierarchical Moderator Analysis Example

Variable	k	Mean	Variance
r_{xx}	5	0.82	0.0026
r_{yy}	3	0.54	0.0051
U	4	0.74	0.0069

Note. r_{xx} = predictor reliability; r_{yy} = criterion reliability; U = range restriction; k = number of data points contributing to artifact distribution; *Mean* = average of artifact distribution; *Variance* = variance of the artifact distribution.

TABLE 3.14
Squared Coefficient of Variation, Compound Attenuation Factor,
and Compound Variance Factor Values for Mental Rotation Test Fully
Hierarchical Moderator Analysis Example

FACTOR	Overall	Female/ Pilots	Female/ Nonpilots	Male/ Pilots	Male/ Nonpilots
SCV(A)	0.00099431	0.00099431	0.00099431	0.00099431	0.00099431
SCV(B)	0.0045898	0.0045898	0.0045898	0.0045898	0.0045898
SCV(C)	0.0089785	0.0075650	0.0063034	0.011320	0.011485
AA	0.50557	0.51448	0.52363	0.49305	0.49225
V	0.014563	0.013149	0.011887	0.016904	0.017070

Note. SCV(A) = squared coefficient of variation for predictor reliability; *SCV(B)* = squared coefficient of variation for criterion reliability; *SCV(C)* = squared coefficient of variation for range restriction; *AA* = compound attenuation factor; *V* = compound variance factor.

With this approach, two primary meta-analytic statistics of interest are the fully corrected mean correlation ($\bar{r}$) and its variance. Thus, unlike the typical Glassian approach to the meta-analysis of effect sizes, the Hunter and Schmidt (1990a) approach to the meta-analysis of correlations corrects for a number of specified statistical artifacts, because it is posited that the observed variability in correlations (the results of studies) is not real, but is instead due to these artifacts.

Specifically, we focused on (a) the calculation of summary statistics, (b) the estimation of sampling error variance, which cumulated in the computation of the percent variance accounted for by sampling error, and (c) the chi-square test for homogeneity and variation in studies, and (d) the computation of confidence intervals to assess the accuracy of the mean sample-weighted correlation. We next discussed (e) the correcting of the summary statistics for multiple artifacts, focusing specifically on the use of artifact distribution to correct for predicator reliability, criterion reliability, and range restriction.

In relation to the chi-square test for homogeneity, we reviewed and discussed (f) the issue of conducting moderator analysis in the meta-analysis of effect sizes. We discussed the logical inconsistency of applying statistical significance tests in the analysis of moderators in meta-analysis. We also noted and acknowledged that despite this, it is not an uncommon practice. In discussing moderator analysis, we (g) drew attention to and highlighted the distinction between confidence intervals and credibility intervals, along with their appropriate use and interpretation.

We also (i) noted the distinction between fully hierarchical and partial or incomplete moderator analysis, and highlighted the advantages of the former—fully hierarchical moderator analysis—over the latter. Although technically superior, fully hierarchical analyses are performed very infrequently in the extant literature. This is because the implementation of a fully hierarchical moderator analysis first calls for the creation of a matrix to represent all the levels of all the moderators. Thus, the feasibility of these analyses is constrained by the number of data points in each cell.

Finally, the above issues were discussed within the context of specific examples which were illustrated with tables presenting the actual SAS code along with representations of the actual SAS printouts. Coupled with these, we presented tables summarizing the meta-analysis results of our examples, along with interpretations of these results. The SAS code and data used in the examples presented in this chapter can be downloaded from www.erlbaum.com. Chapter 4 discusses the issue of outliers in meta-analysis and presents Huffcutt and Arthur's (1995) sample-adjusted meta-analytic deviancy (SAMD) procedure, to date the only outlier technique developed for meta-analytic data.

Outliers in
Meta-Analytic Data

INTRODUCTION AND OVERVIEW

This chapter focuses on the identification of outliers in meta-analytic data. When we use the term *outlier*, we refer to a single data point that is extreme in its value relative to other values of the variable. Outliers are important because they can have a substantial impact on empirical findings, and subsequently, the conclusions that we draw from them. As Cortina and Gully (1999) noted, outliers can exist at many different levels including the data point outlier, the case outlier, the variable outlier, and the *study outlier*. Our primary focus in this chapter is the study outlier. In addition to a general discussion of outliers in meta-analytic data, we present Huffcutt and Arthur's (1995) sample-adjusted meta-analytic deviancy procedure, a technique for identifying outliers in meta-analytic data. The SAS code for conducting this outlier analysis for both correlations (*r*s) and effect sizes (*d*s) is presented along with illustrative examples.

OUTLIERS IN META-ANALYTIC DATA

A number of prominent statisticians have noted that virtually all data sets contain at least some outlier data points (Gulliksen, 1986; Mosteller & Hoaglin, 1991; Tukey, 1960, 1977). An *outlier* is a datum that appears to be inconsistent with the other data of which it is part, due to either error in transcription, computation, or an unusual research participant characteristic. The effect of such outliers is, typically, a notable increase in the observed variance and possibly some distortion of the mean. Some researchers (e.g., Huber, 1980; Tukey, 1960) recommend deleting the most extreme 10% of the values, specifically the highest 5% and the lowest 5%, to make the data set more representative of the population.

In addition to primary studies, it is equally likely that outliers exist with meta-analytic data sets. Schmidt et al. (1993) noted, "Because most validity generalization analyses have been conducted using studies of imperfect methodological quality, the presence of outliers is highly probable" (p. 10). With meta-analysis, the data analyzed are study coefficients (e.g., correlations) rather than individual participant data. Thus, an outlier in the meta-analytic framework would be a primary study result (i.e., correlation, effect size, or other test statistic value) that does not appear to be consistent with the other study results (correlations, effect sizes, or other test statistic values), because of either errors in data collection, computation, or an unusual feature of the study design or the study participants. The effect of such outliers would typically be an increase in the residual variability and a possible shift in the mean correlation or effect size. Thus, it is important to distinguish between looking for outliers within studies (i.e., data point, case, or variable outliers) and looking for outliers among the study correlations or effect sizes, because a given primary study may have no outlier participants yet may still be an outlier in relation to other studies. Orr, Sackett, and DuBois (1991), for example, analyzed the actual participant data from a number of cognitive ability studies and assessed the changes to meta-analytic results after removal of outlier participants within the various studies. They failed, however, to address the issue of outliers among the study correlations.

Detecting outliers in meta-analytic data sets is, potentially, very important. A fundamental premise of meta-analytic procedures is that at least some (and often much) of the observed variation across outcomes of independent studies in a given area results from such statistical artifacts as sampling error (Hunter & Schmidt, 1990a; Hunter et al., 1982). The extent of residual variability after removal of variance accounted for by artifacts is often then used as a basis to make inferences about the relation between the study variables. For example, a high residual variance may suggest that other variables are present that moderate the relation or that validity may not generalize across situations. Thus, it is possible that the presence of outliers in a meta-analytic analysis could alter the conclusions reached.

THE SAMPLE-ADJUSTED META-ANALYTIC DEVIANCY (SAMD) STATISTIC

Although methods for detecting outliers in primary data are numerous and well established, the methodology for detecting outliers in meta-analytic data is much less developed. Traditional outlier techniques such as box plot analysis for univariate data (Mosteller & Hoaglin, 1991) or studentized residuals for bivariate data (Freund & Littell, 1991) are not appropriate because they do not take sample size into account. Such techniques implicitly assume that all data points have equal status. With meta-analytic data, each study coefficient is typically based on a different number of participants and the principle of sampling error (Hunter & Schmidt, 1990a) suggested that study results (correlations or effect sizes), based on smaller samples, are more likely to be deviant than those based on larger samples. Thus, given the wide range of sample sizes typically observed in a meta-analysis, using conventional techniques to identify outlier coefficients may lead to erroneous conclusions. For the same reason subjective visual inspection is also problematic.

To date, there has been only one technique developed specifically for meta-analytic data, the "Sample-Adjusted Meta-Analytic Deviancy" procedure developed by Huffcutt and Arthur (1995). Known more commonly as the SAMD statistic, it compares the value of each study coefficient to the mean sample-weighted coefficient computed without that coefficient in the analysis, then adjusts that difference for the sample size of the study. (Note, for the sake clarity, we use the term *coefficient* to refer to study results expressed as either correlations, effect sizes, or similar test statistics. However, when specifically referring to correlations or effect sizes, they are referred to as such.) The end result is one SAMD statistic for every primary study, and a distribution of SAMD statistics that approximate a *t* distribution. Studies with an extreme SAMD value are then investigated as potential outliers.

The SAMD procedure begins with computation of the mean difference for each primary study, that is, the difference between the individual value of that coefficient and the mean sample-weighted value computed across all of the other studies. Specifically, for each primary study, the sample-weighted mean coefficient is calculated without that study in the analysis, thus ensuring it does not un-

duly influence the result. The difference between the value of each individual coefficient and the sample-weighted mean coefficient computed without the coefficient is a raw measure of deviancy unadjusted for sample size.

Adjusting for sample size can be accomplished by considering the theoretical
sampling distribution for each primary study. When considering a particular
study, the sample-weighted mean coefficient computed without the study represents the best independent estimate of the relation between the variables in the
population (uncorrected for artifacts such as range restriction). One could theoretically take a large number of samples from that population, all the same size as
the primary study, compute the coefficient for each sample and plot the resulting
distribution of coefficients. This distribution represents the variability expected
for the primary study solely based on chance. Equations for computing the variance of the sampling distribution for a particular sample size are shown next, for
correlational (Equation 4.1) and effect size data (Equation 4.2) respectively
(Hunter & Schmidt, 1990a, pp. 105, 281), where i is the ith study in a meta-analysis and N is its sample size.

$$Var(i) = \frac{(1 - \bar{r}^2_{[w/o\ study]})^2}{N - 1} \tag{4.1}$$

$$Var(i) = \frac{4 * (N - 1) * (1 + (\bar{d}^2_{[w/o\ study]} \div 8))}{N(N - 3)} \tag{4.2}$$

The SAS statement necessary to calculate $Var(i)$ for correlations (Equation
4.1) is presented in Table 4.1, program code line 150. Likewise, the SAS program
code for Equation 4.2—$Var(i)$ for effect sizes—is presented in Table 4.3, program code line 130. For both correlational and effect size data, we used the fully
hierarchical moderator analysis hypothetical data presented in the volume (see
Tables 3.10 and 2.10) as illustrative examples. The reader should be aware that all
the SAS code and associated examples presented in this volume were ran on a
mainframe.

Note that the values for the variables listed in *DATA STEP_2* in the programs presented in Table 4.1 (program code lines 139-145) and Table 4.3 (program code lines 119-125) are obtained from the code in *DATA STEP_1* in the respective programs. Thus, the code presented in Tables 4.1 and 4.3 must be run first, and then the specified values for the variables in *DATA STEP_2* inserted and the programs rerun.

In addition, there is likely to be some sampling error in the mean coefficient, which needs to be taken into account as well when assessing the raw deviancy of a particular study. Equations for computing the sampling error variance of a mean coefficient are presented in Equation 4.3 for correlational data, and Equation 4.4 for effect size data (Schmidt, Hunter, & Raju, 1988, p. 668; Hunter & Schmidt, 1990a, p. 427), where N in Equation 4.3 is the total sample size, $\overline{N}$ in Equation 4.4 is the average sample size of the studies, and k in both equations is the number of studies used to compute the mean coefficient.

$$Var(\overline{r}) = \frac{(1 - \overline{r}^2_{[w/o\ study]})^2}{N - k} \tag{4.3}$$

$$Var(\overline{d}) = \frac{4 * (\overline{N} - 1) * (1 + (\overline{d}^2_{[w/o\ study]} \div 8))}{\overline{N} * (\overline{N} - 3) * k} \tag{4.4}$$

The SAS statements necessary to calculate $Var(\overline{r})$—Equation 4.1—are presented in Table 4.1, program code line 151. Similar code for $Var(\overline{d})$—Equation 4.2—can be found in Table 4.3, program code line 132.

These sources of sampling error, in both the study and the mean coefficient, are then combined to form the sampling error of the difference between a study coefficient and the mean coefficient computed without that study. Specifically, the sampling standard error of the difference is the square root of the sum of these two variances.

Finally, the SAMD statistic is computed by dividing the raw deviancy for a given study by the sampling standard error of the difference. The result is a statistic that reflects how deviant a study coefficient is relative to what can be expected from chance for that sample size, and for the mean coefficient with which it is being compared. Dividing by the respective sampling standard errors of the difference across all studies in a meta-analysis results in a distribution of SAMD statistics that roughly approximates a t distribution, thus simplifying interpretation. The final equations for computing the SAMD statistic are presented next for correlational (Equation 4.5) and effect size data (Equation 4.6), respectively.

$$SAMD(i) = \frac{r_{[study\ i]} - \overline{r}_{[w/o\ study]}}{\sqrt{Var(i) + Var(\overline{r})}} \qquad (4.5)$$

$$SAMD(i) = \frac{d_{[study\ i]} - \overline{d}_{[w/o\ study]}}{\sqrt{Var(i) + Var(\overline{d})}} \qquad (4.6)$$

Again, the SAS program code for Equation 4.5 is presented in Table 4.1, program code lines 154-157. Likewise, the code for Equation 4.6 can be found in Table 4.3, program code lines 134-136.

The final step is determining what SAMD values to consider extreme (i.e., as potential outliers). We strongly recommend the use of a scree plot (Dillion & Goldstein, 1984; Loehlin, 1987) to set a cutoff score above which data points are considered to be outliers and further investigated as such. Specifically, the absolute SAMD values are rank ordered from highest to lowest and plotted. SAMD values that rise above the flat gradual slopes are identified as potential outliers. Thus, after having identified specified studies as potential outliers, the researcher should engage in a further assessment of these studies to determine if the deviancy can be attributed to some unusual study feature. As Greenfield (1987) noted, a thorough analysis should be made with all outliers to try and uncover the reason for the apparent deviancy. For instance, if available, the individual participant data for these studies may be scrutinized for computational or transcription error. Based on this information, the decision regarding whether to exclude such studies from a meta-analysis may then be made.

An alternative to the scree plot is a more subjective inspection to find the SAMD values that appear extreme relative to the other values. In the latter case, considering that the SAMD values approximate a t distribution, using a cutoff of approximately 3.0 (and tagging SAMD values greater than 3.0 as potential outliers) would be appropriate.

SAS PRINTOUTS FOR SAMD EXAMPLES AND SAMD SCREE PLOTS

Tables 4.2 and 4.4 present SAS printouts for the program codes presented in Tables 4.1 and 4.3. These printouts include both the SAS program code as well as the output (i.e., results). Figures 4.1 and 4.2 present SAMD scree plots for correlational and effect size examples, respectively. These scree plots show that for both sets of data, a majority of the values formed a somewhat flat gradual line. Indeed, for the correlational data (FIG. 4.1), there appear to be no outliers because there were no drastic breaks in the scree plot. Furthermore, the largest SAMD value was 3.04, a value only marginally larger than the previously discussed alternative 3.00 cutoff rule.

For the effect size data (FIG. 4.2), however, there appear to be two flat linear lines. As Loehlin (1987) noted, such a configuration is common. The first break consists of four data points, and the second, with a slightly steeper slope, consists of only one data point. The SAMD values for these five data points, in descending order, were 4.88, 3.33, 3.18, 2.99, and 2.94. Thus, in practice, these five data points (studies) would most likely be further investigated as potential outliers.

SUMMARY

The SAMD statistic presented in this chapter provides a systematic, standardized technique for identifying studies that do not appear to fit with the other studies in a meta-analysis. When identifying a particular primary study as a potential

TABLE 4.1
SAS Statements for Conducting an SAMD Outlier Analysis of Meta-Analytic Correlational Data

001	DATA OBSERVED;				
002	INPUT STUDY R N SEX $ STATUS $;				
003	RN = R * N;			∗∗∗ WEIGHTING STUDY R BY ITS SAMPLE SIZE	∗∗;
004					
005	CARDS;				
006	1	0.60	85	FEMALE	PILOT
007	2	0.12	70	MALE	NONPILOT
008	3	0.24	125	MALE	PILOT
009	4	0.02	25	MALE	PILOT
010	5	0.50	75	FEMALE	NONPILOT
011	6	0.35	100	FEMALE	NONPILOT
012	7	0.16	75	MALE	NONPILOT
013	8	0.28	125	FEMALE	PILOT
014	9	0.07	50	MALE	PILOT
015	10	0.58	25	FEMALE	NONPILOT
016	11	0.12	97	MALE	NONPILOT
017	12	0.60	9	MALE	PILOT
018	13	0.25	69	MALE	PILOT
019	14	0.17	68	MALE	NONPILOT
020	15	0.57	64	FEMALE	NONPILOT
021	16	0.60	9	MALE	PILOT
022	17	0.34	64	FEMALE	NONPILOT
023	18	0.33	78	FEMALE	NONPILOT
024	19	0.04	87	MALE	PILOT
025	20	0.05	35	MALE	PILOT
026	21	0.51	60	FEMALE	NONPILOT
027	22	0.33	46	FEMALE	NONPILOT
028	23	0.53	70	FEMALE	NONPILOT
029	24	0.05	64	FEMALE	PILOT
030	25	0.59	24	MALE	NONPILOT
031	26	0.19	83	MALE	NONPILOT
032	27	0.18	31	MALE	NONPILOT
033	28	0.43	54	FEMALE	PILOT
034	29	0.60	68	FEMALE	PILOT
035	30	0.15	18	MALE	NONPILOT
036	31	0.43	21	FEMALE	PILOT
037	32	0.26	19	MALE	PILOT
038	33	0.35	34	FEMALE	NONPILOT
039	34	0.19	24	MALE	NONPILOT
040	35	0.30	52	FEMALE	NONPILOT
041	36	0.52	40	FEMALE	NONPILOT
042	37	0.40	87	FEMALE	PILOT
043	38	0.37	29	MALE	NONPILOT
044	39	0.36	35	MALE	NONPILOT
045	40	0.24	42	MALE	PILOT
046	41	0.45	44	FEMALE	PILOT
047	42	0.32	71	FEMALE	NONPILOT
048	43	0.58	75	FEMALE	NONPILOT
049	44	0.58	64	FEMALE	NONPILOT
050	45	0.37	40	FEMALE	NONPILOT
051	46	0.19	34	MALE	NONPILOT
052	47	0.25	19	FEMALE	PILOT
053	48	0.30	15	MALE	NONPILOT
054	49	0.42	51	FEMALE	PILOT
055	50	0.11	37	MALE	NONPILOT
056	51	0.11	35	MALE	NONPILOT
057	52	0.54	85	FEMALE	NONPILOT
058	53	0.19	9	MALE	NONPILOT
059	54	0.58	37	FEMALE	NONPILOT
060	55	0.56	65	FEMALE	NONPILOT

(continued)

```
061    56   0.36  68    FEMALE   NONPILOT
062    57   0.04  28    MALE     PILOT
063    58   0.09  27    MALE     PILOT
064    59   0.32  47    FEMALE   NONPILOT
065    60   0.27  44    FEMALE   PILOT
066    61   0.12  62    MALE     NONPILOT
067    62   0.41  35    FEMALE   PILOT
068    63   0.26  63    FEMALE   PILOT
069    64   0.09  25    MALE     PILOT
070    65   0.29  73    FEMALE   PILOT
071    66   0.13  6     MALE     NONPILOT
072    67   0.36  41    FEMALE   NONPILOT
073    68   0.46  8     FEMALE   PILOT
074    69   0.28  37    FEMALE   PILOT
075    70   0.04  26    MALE     PILOT
076    71   0.33  59    FEMALE   NONPILOT
077    72   0.22  36    MALE     PILOT
078    73   0.02  83    MALE     PILOT
079    74   0.44  89    FEMALE   PILOT
080    75   0.58  16    FEMALE   NONPILOT
081    76   0.28  48    MALE     PILOT
082    77   0.59  7     FEMALE   NONPILOT
083    78   0.44  102   FEMALE   PILOT
084    79   0.07  75    MALE     PILOT
085    80   0.12  49    MALE     NONPILOT
086    81   0.17  46    MALE     NONPILOT
087    82   0.38  52    FEMALE   NONPILOT
088    83   0.27  55    MALE     PILOT
089    84   0.05  62    MALE     PILOT
090    85   0.23  28    MALE     PILOT
091    86   0.44  63    MALE     PILOT
092    87   0.13  58    MALE     NONPILOT
093    88   0.10  21    MALE     NONPILOT
094    89   0.35  40    FEMALE   NONPILOT
095    90   0.17  22    MALE     NONPILOT
096    91   0.21  35    FEMALE   PILOT
097    92   0.36  35    FEMALE   NONPILOT
098    93   0.04  9     MALE     PILOT
099    94   0.28  66    FEMALE   PILOT
100    95   0.57  22    FEMALE   NONPILOT
111    96   0.33  63    FEMALE   NONPILOT
112    97   0.46  59    MALE     PILOT
113    98   0.56  55    FEMALE   NONPILOT
114    99   0.36  75    FEMALE   NONPILOT
115    100  0.12  20    MALE     NONPILOT
116    101  0.21  12    MALE     PILOT
117    102  0.44  40    FEMALE   PILOT
118    103  0.08  56    MALE     PILOT
119    104  0.12  79    MALE     NONPILOT
120    105  0.18  50    MALE     NONPILOT
121    106  0.41  31    FEMALE   PILOT
122    107  0.27  32    FEMALE   PILOT
123    108  0.03  46    MALE     NONPILOT
124    109  0.18  49    FEMALE   NONPILOT
125    110  0.27  5     FEMALE   PILOT
126    ;
127    DATA STEP_1;
128      SET OBSERVED;
129    PROC MEANS MEAN  MAXDEC=2;                    ***** GENERATES "MEAN_R"              ***;
130      VAR R;
131
132    PROC MEANS SUM  MAXDEC=2;
133      VAR RN;                                     ***** GENERATES "TOTAL_R"             ***;
134
135    PROC MEANS MEAN SUM N  MAXDEC=2;
136      VAR N;                                      ***** GENERATES "MEAN_N", "TOTAL_N"   ***;
137                                                  ***** AND "K"                         ***;
138
```

(continued)

```
139    DATA STEP_2;
140      SET STEP_1;
141      MEAN_R = 0.30;
142      TOTAL_R = 1620.41;
143      MEAN_N = 48.70;
144      TOTAL_N = 5357.00;
145      K    = 110;
146
147                                   ••••        COMMENTS                                    ••;
148    TOTAL_NW = TOTAL_N - N;        *TOTAL N (SAMPLE SIZE) WITHOUT_THAT_STUDY               ••;
149    MEAN_RW = (TOTAL_R - RN) / TOTAL_NW;**   MEAN R WITHOUT_THAT_STUDY                     ••;
150    VAR_I = ((1 - (MEAN_RW ** 2)) ** 2) / (N - 1);
151    VAR_R  = ((1 - (MEAN_RW ** 2)) ** 2) / (N - K);
152    IF VAR_I LT 0 THEN VAR_I = 0;
153    IF VAR_R LT 0 THEN VAR_R = 0;
154    SAMD_I = (R - MEAN_RW) / (SQRT (VAR_I + VAR_R));
155
156    SAMD   = ABS (SAMD_I);              ••• GENERATES ABSOLUTE SAMD VALUES                 ••;
157    SAMD   = ROUND (SAMD,.01);          ••• ROUNDING OFF TO 2 DECIMAL PLACES               ••;
158
159    PROC RANK DESCENDING OUT=TEMP;      ••• ASSIGNING RANK ORDER VALUES                    ••;
160      VAR SAMD;                         ••• FOR SAMD, FROM LOWEST TO HIGHEST               ••;
161      RANKS RANK_ID;                    ••• "RANK_ID" IS THE VARIABLE NAME                 ••;
162    PROC SORT;                          ••• FOR THE RANK ORDER                             ••;
163      BY RANK_ID;
164
165    DATA STEP_3;
166      SET TEMP;
167    ••••••••••••••••••••••••••••••••••••••••••••••••••••••••••••••••••••••••••••••••••••••••;
168    ••••••••••••••••••••        WRITING OUT SAMD OUTFILE STARTS HERE                     ••••;
169    FILE OUTPUT;
170      PUT RANK_ID SAMD;
171
172    PROC PRINT;
173      VAR STUDY R SAMD RANK_ID;
174    PROC MEANS MEAN N STD MIN MAX RANGE;
175      VAR SAMD;
```

Note. The values for the variables listed in *DATA STEP_2* in the programs presented in Table 4.1 (program code lines 139-145) are obtained from the code in *DATA STEP_1* (program code line 127-137). Thus, the code presented in Table 4.1 must be run first, and the specified values for the variables in *DATA STEP_2* inserted and the program rerun. The program code presented in Table 4.1 also writes out an outfile of SAMD values (program code lines 169-170). To accomplish this, depending on the computer system, it may be necessary to place certain job control language (JCL) statements before the SAS statements.

Program code lines 152 and 153 set *VAR_I* and *VAR_R* to zero (0) if they are less than zero (i.e., if they have negative values, as was the case in our specific example). This is necessary to ensure that the computation of the square-root of the sum of these coefficients (program code line 154) is not performed on a negative value, which would be an invalid mathematical operation.

TABLE 4.2
SAS Printout for SAMD Outlier Analysis of Meta-Analytic Correlational Data

```
1                          The SAS System          03:43 Saturday, May 27, 2000

NOTE: Copyright (c) 1989-1996 by SAS Institute Inc., Cary, NC, USA.
NOTE: SAS (r) Proprietary Software Release 6.09  TS470

****************************************************************
*                                                              *
*      Welcome to the new SAS System, Release 6.09 Enhanced    *
*                                                              *
****************************************************************

NOTE: SAS system options specified are:
      SORT=4

NOTE: The initialization phase used 0.19 CPU seconds and 2041K.
1          OPTIONS LS=90;
2          DATA OBSERVED;
3           INPUT STUDY R  N  SEX $  STATUS $;
4          RN = R * N;  *** WEIGHTING STUDY R BY ITS SAMPLE SIZE  **;
5
6           CARDS;

NOTE: The data set WORK.OBSERVED has 110 observations and 6 variables.
NOTE: The DATA statement used 0.07 CPU seconds and 2679K.

117         ;
118
119         DATA STEP_1;
120           SET OBSERVED;

NOTE: The data set WORK.STEP_1 has 110 observations and 6 variables.
NOTE: The DATA statement used 0.01 CPU seconds and 2683K.

121         PROC MEANS MEAN  MAXDEC=2;
122           VAR R;          *****  GENERATES "MEAN_R"          **;
123

NOTE: The PROCEDURE MEANS printed page 1.
NOTE: The PROCEDURE MEANS used 0.03 CPU seconds and 2888K.

124         PROC MEANS SUM  MAXDEC=2;
125           VAR RN;         *****  GENERATES "TOTAL_R"         **;
126

NOTE: The PROCEDURE MEANS printed page 2.
NOTE: The PROCEDURE MEANS used 0.01 CPU seconds and 2888K.

2                          The SAS System          03:43 Saturday, May 27, 2000

127         PROC MEANS MEAN SUM N  MAXDEC=2;
128           VAR N;          *****  GENERATES "MEAN_N",  "TOTAL_N"  **;
129                           *****  AND   "K"                  **;
130

NOTE: The PROCEDURE MEANS printed page 3.
NOTE: The PROCEDURE MEANS used 0.01 CPU seconds and 2888K.

131         DATA STEP_2;
132           SET STEP_1;
133         MEAN_R  = 0.30;
134         TOTAL_R = 1620.41;
135         MEAN_N  = 48.70;
136         TOTAL_N = 5357.00;
137         K       = 110;
```

(continued)

```
138                         **********************************************;
139                         **************** COMMENTS ******************;
140       TOTAL_NW = TOTAL_N - N; * TOTAL N (SAMPLE SIZE) WITHOUT _THAT_ STUDY *;
141       MEAN_RW = (TOTAL_R - RN) / TOTAL_NW; ** MEAN R WITHOUT _THAT_ STUDY **;
142       VAR_I   = ((1 - (MEAN_RW ** 2)) ** 2) / (N - 1);
143       VAR_R   = ((1 - (MEAN_RW ** 2)) ** 2) / (N - K);
144       IF VAR_I LT 0  THEN VAR_I=0;
145       IF VAR_R LT 0  THEN VAR_R=0;
146       SAMD_I  = (R - MEAN_RW) / (SQRT (VAR_I + VAR_R));
147
148       SAMD   = ABS (SAMD_I);   **** GENERATES ABSOLUTE SAMD VALUES    ***;
149       SAMD   = ROUND (SAMD,.01); *** ROUNDING OFF TO 2 DECIMAL PLACES ***;
150

NOTE: The data set WORK.STEP_2 has 110 observations and 17 variables.
NOTE: The DATA statement used 0.05 CPU seconds and 2977K.

151       PROC RANK DESCENDING OUT=TEMP; **  ASSIGNING RANK ORDER VALUES      **;
152         VAR SAMD;                    ***  FOR SAMD, FROM LOWEST TO HIGHEST **;
153         RANKS RANK_ID;               ***  "RANK_ID" IS THE VARIABLE NAME   **;

NOTE: The data set WORK.TEMP has 110 observations and 18 variables.
NOTE: The PROCEDURE RANK used 0.01 CPU seconds and 2962K.

154       PROC SORT;                       ***  FOR THE RANK ORDER            **;
155         BY RANK_ID;
156

NOTE: The data set WORK.TEMP has 110 observations and 18 variables.
NOTE: The PROCEDURE SORT used 0.02 CPU seconds and 3166K.

157       DATA STEP_3;
158         SET TEMP;
159       ****************************************************************;
160       ******************** WRITING OUT SAMD OUTFILE STARTS HERE *********;
161       FILE OUTPUT;
162         PUT  RANK_ID SAMD;
163

NOTE: The file OUTPUT is:
      Dsname=E021093.META.ANALYSIS.ROUT,
      Unit=3380,Volume=USR005,Disp=NEW,Blksize=6320,
      Lrecl=80,Recfm=FB
3                                   The SAS System        03:43 Saturday, May 27, 2000

NOTE: 110 records were written to the file OUTPUT.
NOTE: The data set WORK.STEP_3 has 110 observations and 18 variables.
NOTE: The DATA statement used 0.06 CPU seconds and 3198K.

164       PROC PRINT;
165         VAR  STUDY  R  SAMD RANK_ID;
NOTE: The PROCEDURE PRINT printed pages 4-5.
NOTE: The PROCEDURE PRINT used 0.03 CPU seconds and 3281K.

166       PROC MEANS MEAN N STD MIN MAX RANGE;
167         VAR SAMD;
NOTE: The PROCEDURE MEANS printed page 6.
NOTE: The PROCEDURE MEANS used 0.01 CPU seconds and 3281K.

NOTE: The SAS session used 0.53 CPU seconds and 3281K.
NOTE: SAS Institute Inc., SAS Campus Drive, Cary, NC USA 27513-2414
```

(continued)

```
                        The SAS System        03:43 Saturday, May 27, 2000   1

                        Analysis Variable : R

                                      Mean
                              ------------
                                      0.30
                              ------------
                        The SAS System        03:43 Saturday, May 27, 2000   2

                        Analysis Variable : RN

                                      Sum
                              ------------
                                  1620.41
                              ------------
                        The SAS System        03:43 Saturday, May 27, 2000   3

                        Analysis Variable : N

                              Mean          Sum      N
                        -------------------------------
                              48.70      5357.00    110
                        -------------------------------
                        The SAS System        03:43 Saturday, May 27, 2000   4

            OBS    STUDY      R      SAMD     RANK_ID

             1        1     0.60     3.04       1.0
             2       73     0.02     2.87       2.0
             3       19     0.04     2.73       3.0
             4       29     0.60     2.71       4.0
             5       43     0.58     2.66       5.0
             6       44     0.58     2.45       6.0
             7       52     0.54     2.43       7.0
             8       15     0.57     2.36       8.0
             9       55     0.56     2.29       9.0
            10       24     0.05     2.24      10.5
            11       79     0.07     2.24      10.5
            12       84     0.05     2.20      12.0
            13       23     0.53     2.10      13.5
            14       98     0.56     2.10      13.5
            15      108     0.03     2.03      15.0
            16       11     0.12     2.01      16.0
            17        5     0.50     1.89      17.0
            18       54     0.58     1.84      18.5
            19      103     0.08     1.84      18.5
            20        9     0.07     1.81      20.0
            21      104     0.12     1.80      21.0
            22       21     0.51     1.77      22.0
            23        2     0.12     1.69      23.0
            24       20     0.05     1.63      24.0
            25       61     0.12     1.59      25.0
            26       78     0.44     1.55      26.0
            27        4     0.02     1.53      27.0
            28       25     0.59     1.52      28.0
            29       57     0.04     1.51      29.0
            30       10     0.58     1.50      30.5
            31       36     0.52     1.50      30.5
            32       70     0.04     1.45      32.5
            33       87     0.13     1.45      32.5
            34       74     0.44     1.44      34.0
            35       80     0.12     1.41      35.0
```

(continued)

36	7	0.16	1.37	36.0
37	95	0.57	1.35	37.0
38	97	0.46	1.33	38.0
39	50	0.11	1.28	39.0
40	51	0.11	1.24	40.0
41	14	0.17	1.21	41.0
42	58	0.09	1.20	42.5
43	86	0.44	1.20	42.5
44	75	0.58	1.19	44.0
45	64	0.09	1.15	45.0
46	26	0.19	1.14	46.0
47	41	0.45	1.07	47.0
48	28	0.43	1.03	48.0
49	37	0.40	1.01	49.0
50	88	0.10	1.00	50.0
51	81	0.17	0.99	51.0
52	102	0.44	0.95	52.5
53	105	0.18	0.95	52.5
54	109	0.18	0.94	54.0
55	12	0.60	0.93	55.5
56	16	0.60	0.93	55.5

The SAS System   03:43 Saturday, May 27, 2000   5

OBS	STUDY	R	SAMD	RANK_ID
57	49	0.42	0.92	57.0
58	100	0.12	0.88	58.0
59	93	0.04	0.82	59.0
60	77	0.59	0.78	60.0
61	27	0.18	0.74	61.0
62	46	0.19	0.72	62.0
63	30	0.15	0.69	63.5
64	62	0.41	0.69	63.5
65	90	0.17	0.67	65.0
66	106	0.41	0.65	66.0
67	31	0.43	0.63	67.0
68	82	0.38	0.61	68.0
69	34	0.19	0.60	69.5
70	91	0.21	0.60	69.5
71	99	0.36	0.55	71.0
72	72	0.22	0.54	72.0
73	6	0.35	0.53	73.0
74	56	0.36	0.52	74.0
75	13	0.25	0.48	75.0
76	45	0.37	0.47	76.0
77	68	0.46	0.46	77.0
78	40	0.24	0.44	78.0
79	66	0.13	0.43	79.0
80	85	0.23	0.42	80.0
81	38	0.37	0.40	81.5
82	67	0.36	0.40	81.5
83	39	0.36	0.37	84.0
84	63	0.26	0.37	84.0
85	92	0.36	0.37	84.0
86	53	0.19	0.35	86.0
87	101	0.21	0.34	87.0
88	17	0.34	0.33	88.5
89	89	0.35	0.33	88.5
90	33	0.35	0.30	90.0
91	18	0.33	0.27	91.5
92	83	0.27	0.27	91.5
93	3	0.24	0.26	93.0
94	47	0.25	0.25	94.0
95	60	0.27	0.24	95.5
96	96	0.33	0.24	95.5
97	71	0.33	0.23	97.0
98	22	0.33	0.20	99.5
99	32	0.26	0.20	99.5
100	94	0.28	0.20	99.5

(continued)

```
                101    107    0.27    0.20      99.5
                102     76    0.28    0.17     102.0
                103     42    0.32    0.16     103.0
                104     69    0.28    0.15     104.0
                105     59    0.32    0.13     105.0
                106     65    0.29    0.12     106.0
                107      8    0.28    0.09     107.0
                108    110    0.27    0.07     108.0
                109     35    0.30    0.02     109.0
                110     48    0.30    0.01     110.0
                          The SAS System        03:43 Saturday, May 27, 2000   6

Analysis Variable : SAMD


       Mean     N      Std Dev      Minimum       Maximum        Range
   --------------------------------------------------------------------------
    1.0410909   110    0.7610559    0.0100000     3.0400000     3.0300000
   --------------------------------------------------------------------------
```

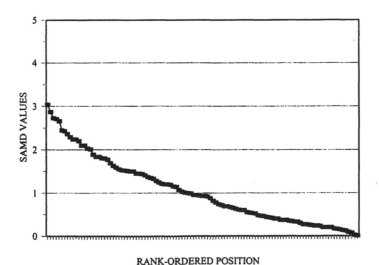

FIG. 4.1. Scree plot analysis for SAMD statistics for correlational data and example presented in Tables 4.1 and 4.2.

TABLE 4.3
SAS Statements for Conducting an SAMD Outlier Analysis of Meta-Analytic Effect Size Data

001	DATA OBSERVED;
002	INPUT STUDY D N TASK $ TEST $;
003	DN = D * N; *** WEIGHTING STUDY D BY ITS SAMPLE SIZE **;
004	CARDS;
005	1 0.20 15 ACCURACY RECALL
006	2 0.32 140 ACCURACY RECALL
007	3 0.88 145 SPEED RECOGN
008	4 0.95 35 SPEED RECOGN
009	5 0.51 45 SPEED RECOGN
010	6 0.43 135 ACCURACY RECOGN
011	7 0.08 180 ACCURACY RECALL
012	8 0.01 50 SPEED RECALL
013	9 0.17 65 ACCURACY RECOGN
014	10 0.50 70 SPEED RECALL
015	11 0.14 80 ACCURACY RECALL
016	12 0.05 135 ACCURACY RECOGN
017	13 0.16 15 SPEED RECOGN
018	14 0.14 120 SPEED RECALL
019	15 0.61 60 ACCURACY RECOGN
020	16 0.08 130 SPEED RECALL
021	17 0.99 230 SPEED RECOGN
022	18 0.21 90 SPEED RECALL
023	19 0.16 180 SPEED RECALL
024	20 0.74 225 ACCURACY RECOGN
025	21 0.80 98 SPEED RECOGN
026	22 0.42 80 SPEED RECOGN
027	23 0.84 15 SPEED RECOGN
028	24 0.07 215 ACCURACY RECALL
029	25 0.37 35 SPEED RECALL
030	26 0.99 110 SPEED RECOGN
031	27 0.53 110 SPEED RECALL
032	28 0.56 40 ACCURACY RECOGN
033	29 0.51 240 SPEED RECOGN
034	30 0.04 115 SPEED RECOGN
035	31 0.34 50 SPEED RECOGN
036	32 0.26 320 ACCURACY RECALL
037	33 0.62 45 SPEED RECALL
038	34 0.82 10 SPEED RECOGN
039	35 0.42 10 SPEED RECOGN
040	36 0.56 35 SPEED RECOGN
041	37 0.86 25 SPEED RECOGN
042	38 0.13 20 SPEED RECALL
043	39 0.43 40 SPEED RECOGN
044	40 0.50 45 ACCURACY RECOGN
045	41 0.33 220 ACCURACY RECOGN
046	42 0.24 20 ACCURACY RECOGN
047	43 0.22 45 SPEED RECALL
048	44 0.14 20 ACCURACY RECALL
049	45 0.27 20 SPEED RECALL
050	46 0.32 80 ACCURACY RECOGN
051	47 0.38 100 SPEED RECOGN

(continued)

052	48	0.33	120	SPEED	RECALL
053	49	0.16	20	ACCURACY	RECALL
054	50	0.69	15	SPEED	RECALL
055	51	0.89	15	SPEED	RECOGN
056	52	0.85	60	SPEED	RECOGN
057	53	0.46	55	SPEED	RECALL
058	54	0.40	20	SPEED	RECALL
059	55	0.34	15	SPEED	RECALL
060	56	0.05	120	ACCURACY	RECALL
061	57	0.95	15	SPEED	RECOGN
062	58	0.66	75	SPEED	RECOGN
063	59	0.09	30	ACCURACY	RECALL
064	60	0.18	60	ACCURACY	RECALL
065	61	0.05	110	SPEED	RECALL
066	62	0.20	155	SPEED	RECOGN
067	63	0.57	90	SPEED	RECOGN
068	64	0.10	95	ACCURACY	RECALL
069	65	0.27	105	SPEED	RECOGN
070	66	0.15	20	SPEED	RECOGN
071	67	0.02	150	ACCURACY	RECALL
072	68	0.09	80	ACCURACY	RECALL
073	69	0.09	100	ACCURACY	RECOGN
074	70	0.16	45	SPEED	RECALL
075	71	0.53	80	ACCURACY	RECOGN
076	72	0.38	75	SPEED	RECALL
077	73	0.29	25	SPEED	RECALL
078	74	0.51	50	ACCURACY	RECOGN
079	75	0.34	105	SPEED	RECOGN
080	76	0.06	25	ACCURACY	RECALL
081	77	0.04	20	ACCURACY	RECOGN
082	78	0.24	40	ACCURACY	RECALL
083	79	0.70	70	SPEED	RECOGN
084	80	0.10	25	ACCURACY	RECOGN
085	81	0.56	20	SPEED	RECOGN
086	82	0.66	60	SPEED	RECOGN
087	83	0.81	40	SPEED	RECOGN
088	84	0.16	90	ACCURACY	RECALL
089	85	0.48	80	SPEED	RECOGN
090	86	0.35	115	ACCURACY	RECOGN
091	87	0.00	110	SPEED	RECALL
092	88	0.11	190	SPEED	RECOGN
093	89	0.07	220	ACCURACY	RECALL
094	90	0.59	35	SPEED	RECALL
095	91	0.97	95	SPEED	RECOGN
096	92	0.91	20	SPEED	RECOGN
097	93	0.12	25	ACCURACY	RECALL
098	94	0.03	95	ACCURACY	RECALL
099	95	0.97	40	ACCURACY	RECOGN
100	96	0.69	80	SPEED	RECOGN
101	97	0.27	15	SPEED	RECALL

(continued)

```
102    98    0.10   260   ACCURACY      RECALL
103    99    0.20   135   SPEED         RECALL
104    100   0.67   15    SPEED         RECOGN
105    ;
106
107    DATA STEP_1;
108      SET OBSERVED;
109    PROC MEANS MEAN MAXDEC=2;
110      VAR D;                              ***** GENERATES "MEAN_D"                    **;
111
112    PROC MEANS SUM MAXDEC=2;
113      VAR DN;                             ***** GENERATES "TOTAL_D"                   **;
114
115    PROC MEANS MEAN SUM N MAXDEC=2;
116      VAR N;                              ***** GENERATES "MEAN_N", "TOTAL_N"         **;
117                                          ***** AND "K"                       N      **;
118
119    DATA STEP_2;
120      SET STEP_1;
121      MEAN_D = 0.39;
122      TOTAL_D = 2836.30;
123      MEAN_N = 81.33;
124      TOTAL_N = 8133.00;
125      K     = 100;
126                                          ***********************************************;
127                                          ***          COMMENTS             ...       **;
128    TOTAL_NW = TOTAL_N - N;               ** TOTAL N (SAMPLE SIZE) WITHOUT _THAT_ STUDY **;
129    MEAN_DW = (TOTAL_D - DN) / TOTAL_NW;  *** MEAN D WITHOUT _THAT_ STUDY       si .   **;
130    VAR_I  = (4 * (N - 1) * (1 + ((MEAN_DW ** 2) / 8))) /
131             (N * (N - 3));
132    VAR_D  = (4 * (MEAN_N - 1) * (1 + ((MEAN_DW ** 2) / 8))) /
133             (MEAN_N * (MEAN_N - 3) * K);
134    SAMD_I = (D - MEAN_DW) / (SQRT (VAR_I + VAR_D));
135    SAMD   = ABS (SAMD_I);                *** GENERATES ABSOLUTE SAMD VALUES         **;
136    SAMD   = ROUND (SAMD,.01);            *** ROUNDING OFF TO 2 DECIMAL PLACES       **;
137
138    PROC RANK DESCENDING OUT=TEMP;        *** ASSIGNING RANK ORDER VALUES           **;
139      VAR SAMD;                           *** FOR SAMD, FROM LOWEST TO HIGHEST      **;
140      RANKS RANK_ID;                      *** "RANK_ID" IS THE VARIABLE NAME        **;
141    PROC SORT;                            *** FOR THE RANK ORDER                    **;
142      BY RANK_ID;
143
144    DATA STEP_3;
145      SET TEMP;
146    *********************************************************************************;
147    **************************   WRITING OUT SAMD OUTFILE STARTS HERE   ***********;
148    FILE OUTPUT;
149      PUT RANK_ID SAMD;
150
151    PROC PRINT;
152      VAR STUDY D SAMD RANK_ID;
153    PROC MEANS MEAN N STD MIN MAX RANGE;
154      VAR SAMD;
```

Note. The values for the variables listed in *DATA STEP_2* in the programs presented in Table 4.3 (program code lines 119-125) are obtained from the code in *DATA STEP_1* (program code line 107-117). Thus, the code presented in Table 4.3 must be run first, and then the specified values for the variables in *DATA STEP_2* inserted and the program rerun. The program code presented in Table 4.3 also writes out an outfile of SAMD values (program code lines 148-149). To accomplish this, depending on the computer system, it may be necessary to place certain job control language (JCL) statements before the SAS statements.

TABLE 4.4
SAS Printout for SAMD Outlier Analysis of Meta-Analytic Effect Size Data

```
1                          The SAS System        03:45 Saturday, May 27, 2000

NOTE: Copyright (c) 1989-1996 by SAS Institute Inc., Cary, NC, USA.
NOTE: SAS (r) Proprietary Software Release 6.09  TS470

•••••••••••••••••••••••••••••••••••••••••••••••••••••••••••••••••••••••
•                                                                     •
•          Welcome to the new SAS System, Release 6.09 Enhanced       •
•                                                                     •
•••••••••••••••••••••••••••••••••••••••••••••••••••••••••••••••••••••••

NOTE: SAS system options specified are:
      SORT=4

NOTE: The initialization phase used 0.18 CPU seconds and 2041K.
1          OPTIONS LS=90;
2          DATA OBSERVED;
3            INPUT STUDY D  N   TASK $  TEST $;
4            DN = D * N;           *** WEIGHTING STUDY D BY ITS SAMPLE SIZE  **;
5            CARDS;

NOTE: The data set WORK.OBSERVED has 100 observations and 6 variables.
NOTE: The DATA statement used 0.07 CPU seconds and 2679K.

106        ;
107
108        DATA STEP_1;
109          SET OBSERVED;

NOTE: The data set WORK.STEP_1 has 100 observations and 6 variables.
NOTE: The DATA statement used 0.01 CPU seconds and 2683K.

110        PROC MEANS MEAN  MAXDEC=2;
111          VAR D;              *****  GENERATES "MEAN_D"           **;
112

NOTE: The PROCEDURE MEANS printed page 1.
NOTE: The PROCEDURE MEANS used 0.03 CPU seconds and 2888K.

113        PROC MEANS SUM  MAXDEC=2;
114          VAR DN;             *****  GENERATES "TOTAL_D"          **;
115

NOTE: The PROCEDURE MEANS printed page 2.
NOTE: The PROCEDURE MEANS used 0.01 CPU seconds and 2888K.

116        PROC MEANS MEAN SUM N  MAXDEC=2;
2                          The SAS System        03:45 Saturday, May 27, 2000

117          VAR N;              *****  GENERATES "MEAN_N",  "TOTAL_N"  **;
118                              *****  AND  "K"                    **;
119

NOTE: The PROCEDURE MEANS printed page 3.
NOTE: The PROCEDURE MEANS used 0.01 CPU seconds and 2888K.

120        DATA STEP_2;
121          SET STEP_1;
122          MEAN_D  = 0.39;
123          TOTAL_D = 2836.30;
124          MEAN_N  = 81.33;
125          TOTAL_N = 8133.00;
126          K       = 100;
127                              ***********************************************;
128                              **************  COMMENTS  ******************;
129          TOTAL_NW = TOTAL_N - N;  * TOTAL N (SAMPLE SIZE) WITHOUT _THAT_ STUDY *;
130          MEAN_DW = (TOTAL_D - DN) / TOTAL_NW;  ** MEAN D WITHOUT _THAT_ STUDY **;
131          VAR_I   = (4 * (N - 1) * (1 + ((MEAN_DW ** 2) / 8))) /
132                    (N * (N - 3));
```

(continued)

```
133        VAR_D    = (4 * (MEAN_N - 1) * (1 + ((MEAN_DW ** 2) / 8))) /
134                   (MEAN_N * (MEAN_N - 3) * K);
135        SAMD_I   = (D - MEAN_DW) / (SQRT (VAR_I + VAR_D));
136        SAMD     = ABS (SAMD_I);    **** GENERATES ABSOLUTE SAMD VALUES    ***;
137        SAMD     = ROUND (SAMD,.01); *** ROUNDING OFF TO 2 DECIMAL PLACES  ***;
138
```

NOTE: The data set WORK.STEP_2 has 100 observations and 17 variables.
NOTE: The DATA statement used 0.05 CPU seconds and 2977K.

```
139        PROC RANK DESCENDING OUT=TEMP; **  ASSIGNING RANK ORDER VALUES      **;
140           VAR SAMD;                    *** FOR SAMD, FROM LOWEST TO HIGHEST **;
141           RANKS RANK_ID;               *** "RANK_ID" IS THE VARIABLE NAME   **;
```

NOTE: The data set WORK.TEMP has 100 observations and 18 variables.
NOTE: The PROCEDURE RANK used 0.01 CPU seconds and 2962K.

```
142        PROC SORT;                      *** FOR THE RANK ORDER               **;
143           BY RANK_ID;
144
```

NOTE: The data set WORK.TEMP has 100 observations and 18 variables.
NOTE: The PROCEDURE SORT used 0.02 CPU seconds and 3166K.

```
145        DATA STEP_3;
146           SET TEMP;
147        *******************************************************************;
148        ******************** WRITING OUT SAMD OUTFILE STARTS HERE *********;
149        FILE OUTPUT;
150           PUT RANK_ID SAMD;
151
```

NOTE: The file OUTPUT is:
      Dsname=I021092.META.ANALYSIS.DOUT,
      Unit=3380,Volume=USR003,Disp=NEW,Blksize=6320,
      Lrecl=80,Recfm=FB

NOTE: 100 records were written to the file OUTPUT.
3   The SAS System   03:45 Saturday, May 27, 2000

NOTE: The data set WORK.STEP_3 has 100 observations and 18 variables.
NOTE: The DATA statement used 0.06 CPU seconds and 3198K.

```
152        PROC PRINT;
153           VAR  STUDY  D  SAMD RANK_ID;
```

NOTE: The PROCEDURE PRINT printed pages 4-5.
NOTE: The PROCEDURE PRINT used 0.02 CPU seconds and 3281K.

```
154        PROC MEANS MEAN N STD MIN MAX RANGE;
155           VAR SAMD;
```
NOTE: The PROCEDURE MEANS printed page 6.
NOTE: The PROCEDURE MEANS used 0.01 CPU seconds and 3281K.

NOTE: The SAS session used 0.52 CPU seconds and 3281K.
NOTE: SAS Institute Inc., SAS Campus Drive, Cary, NC USA 27513-2414
                          The SAS System   03:45 Saturday, May 27, 2000   1

                          Analysis Variable : D

                                     Mean

                                     0.39

(continued)

```
                    The SAS System        03:45 Saturday, May 27, 2000   2

                    Analysis Variable : DN

                              Sum
                           -----------
                            2836.30
                           -----------
                    The SAS System        03:45 Saturday, May 27, 2000   3

                    Analysis Variable : N

                    Mean           Sum    N
                ---------------------------------
                    81.33        8133.00  100
                ---------------------------------
                    The SAS System        03:45 Saturday, May 27, 2000   4
```

OBS	STUDY	D	SAMD	RANK_ID
1	17	0.99	4.88	1.0
2	26	0.99	3.33	2.0
3	3	0.88	3.18	3.0
4	91	0.97	2.99	4.0
5	20	0.74	2.94	5.0
6	21	0.80	2.21	6.0
7	89	0.07	2.07	7.0
8	24	0.07	2.05	8.0
9	98	0.10	2.01	9.0
10	67	0.02	2.00	10.0
11	52	0.85	1.90	11.5
12	95	0.97	1.90	11.5
13	87	0.00	1.81	13.0
14	7	0.08	1.80	14.0
15	4	0.95	1.72	15.5
16	12	0.05	1.72	15.5
17	30	0.04	1.64	17.5
18	88	0.11	1.64	17.5
19	56	0.05	1.62	19.0
20	61	0.05	1.55	20.0
21	94	0.03	1.53	21.0
22	16	0.08	1.52	22.0
23	96	0.69	1.50	23.0
24	79	0.70	1.44	24.0
25	83	0.81	1.41	25.0
26	58	0.66	1.33	26.0
27	69	0.09	1.28	27.0
28	19	0.16	1.26	28.0
29	29	0.51	1.25	29.0
30	37	0.86	1.22	30.0
31	64	0.10	1.20	31.0
32	82	0.66	1.18	32.5
33	92	0.91	1.18	32.5
34	8	0.01	1.17	34.0
35	68	0.09	1.14	35.0
36	14	0.14	1.13	36.0
37	57	0.95	1.07	37.0
38	63	0.57	1.04	38.0
39	15	0.61	0.99	39.0
40	51	0.89	0.96	40.0
41	27	0.53	0.94	41.0
42	11	0.14	0.92	42.5
43	62	0.20	0.92	42.5
44	23	0.84	0.88	45.0
45	33	0.62	0.88	45.0

(continued)

46	84	0.16	0.88	45.0
47	99	0.20	0.86	47.0
48	32	0.26	0.80	48.5
49	71	0.53	0.80	48.5
50	9	0.17	0.71	50.0
51	76	0.06	0.69	51.5
52	90	0.59	0.69	51.5
53	59	0.09	0.68	53.0
54	18	0.21	0.65	55.5
55	28	0.56	0.65	55.5
56	34	0.82	0.65	55.5

The SAS System   03:45 Saturday, May 27, 2000   5

OBS	STUDY	D	SAMD	RANK_ID
57	77	0.04	0.65	55.5
58	60	0.18	0.64	58.0
59	10	0.50	0.62	59.5
60	70	0.16	0.62	59.5
61	50	0.69	0.61	61.0
62	36	0.56	0.60	62.0
63	80	0.10	0.59	63.0
64	85	0.48	0.58	64.0
65	100	0.67	0.57	65.0
66	74	0.51	0.56	66.0
67	93	0.12	0.54	67.0
68	5	0.51	0.53	68.0
69	40	0.50	0.49	69.0
70	6	0.43	0.47	70.0
71	38	0.13	0.46	71.0
72	44	0.14	0.44	72.5
73	81	0.56	0.44	72.5
74	43	0.22	0.42	74.5
75	66	0.15	0.42	74.5
76	49	0.16	0.40	77.0
77	53	0.46	0.40	77.0
78	65	0.27	0.40	77.0
79	13	0.16	0.34	79.0
80	78	0.24	0.33	80.0
81	22	0.42	0.31	81.0
82	1	0.20	0.26	82.0
83	39	0.43	0.25	83.0
84	42	0.24	0.23	84.0
85	2	0.32	0.17	85.5
86	45	0.27	0.17	85.5
87	47	0.38	0.15	87.0
88	41	0.33	0.14	89.0
89	73	0.29	0.14	89.0
90	97	0.27	0.14	89.0
91	46	0.32	0.13	91.5
92	72	0.38	0.13	91.5
93	54	0.40	0.11	93.0
94	35	0.42	0.10	94.5
95	48	0.33	0.10	94.5
96	25	0.37	0.06	96.0
97	75	0.34	0.04	97.0
98	31	0.34	0.03	98.0
99	55	0.34	0.02	99.0
100	86	0.35	0.01	100.0

The SAS System   03:45 Saturday, May 27, 2000   6

Analysis Variable : SAMD

Mean	N	Std Dev	Minimum	Maximum	Range
0.9717000	100	0.8281585	0.0100000	4.8800000	4.8700000

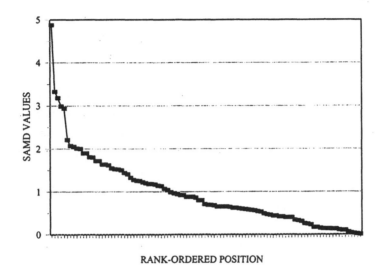

FIG. 4.2. Scree plot analysis for SAMD statistics for effect size data and example presented in Tables 4.3 and 4.4.

outlier, an attempt should be made to ascertain the reason why. However, this is usually difficult because most primary studies report only summary statistics, not raw data. Thus, inevitably, no apparent reason will be found for some (and possibly many) of the studies identified as extreme. In these cases the deviancy may be because of unverifiable transcription or computation error, the presence of some unknown moderator variable operating on a very limited number of studies, or unusually high (but unreported) levels of artifacts such as measurement error (particularly in the criterion) or range restriction. Alternately, the deviancy may be due to extreme sampling error, that is, the large variations which occasionally occur by chance when randomly sampling from a population.

In the absence of an identifiable overt cause, whether to exclude extreme studies or not is an interesting issue. Unlike most methods for analyzing primary data, meta-analysis focuses on estimation of the true variability (i.e., standard deviation) of a relationship in the population. The formula for computing sampling error variance in meta-analysis allows for deviations due to sampling error at their expected frequency, and thus, eliminating coefficients where the deviancy is due entirely to extreme sampling error may result in an underestimation of the true variability. On the other hand, failure to eliminate studies where the deviancy is due to other causes such as transcription error may lead to overestimation. Unfortunately, there is no way of knowing which of these studies are true outliers.

Thus, the issue of whether to exclude extreme studies with no identifiable cause appears to represent a trade-off. Eliminating all such studies may lead to underestimation of the true variability while failure to do so may result in overestimation. However, since the probability of extreme sampling error is very low, it would appear that the underestimation caused by the occasional elimination of a study that is not really an outlier in most cases will be much smaller than the overestimation that would result from not eliminating these studies.

It is also interesting to note that researchers in many of the hard sciences such as chemistry and physics routinely make extensive use of outlier analysis and often eliminate 40% or more of the data, while researchers in the social and behavioral sciences appear much more reluctant to eliminate any data (Hedges, 1987). Nevertheless, we caution that the routine removal of extreme studies be done on a limited basis. The danger with eliminating too many primary studies is that they may actually represent some unknown moderator variable. Indeed, the best decisions to eliminate studies as outliers are those that can be justified both empirically and substantively (Cortina & Gully, 1999; Huffcutt & Arthur, 1995).

We conclude this chapter by noting that the equations presented here apply to meta-analyses where study coefficients are *not* individually corrected for artifacts such as measurement error and range restriction (Hunter & Schmidt, 1990a). Differences between studies in the levels of such artifacts are not taken into account; and although these sources of variance are typically small, they do increase the probability that some values may be incorrectly identified as outliers. Correcting each study coefficient individually for artifacts eliminates this problem, assuming of course that complete artifact data is available for every study. Alternately, the variance due to these artifacts (see Hunter & Schmidt, 1990a) could be added to the denominator of the SAMD formula as an additional term along with the variance of the study coefficient and the variance of the mean coefficient. We now briefly present instructions for modifying the SAMD equations for use with individually corrected coefficients.

Computing SAMD Values with Individually Corrected Correlations

As Hunter and Schmidt (1990a, pp. 145, 146) noted, the sampling error variance of a corrected study is computed in two steps. First, the sampling error variance of the study is computed using the mean uncorrected correlation (Equation 4.1, this chapter). (Note that the mean uncorrected correlation should be computed using the product of the sample size and the square of the compound attenuation factor as a weight rather than simply the sample size; see Hunter & Schmidt, p. 148.) Second, the sampling error variance computed in the first step is divided by the square of the compound attenuation factor, resulting in the sampling error variance of the corrected correlation. Similarly, the sampling error variance in the mean coefficient computed from Equation 4.3 should be corrected by dividing it by the square of the compound attenuation factor as well. The corrected values of the sampling error variance for the study and for the mean coefficient are then used in the denominator of Equation 4.5 to compute the SAMD statistic whereas the corrected values of the study correlation and the mean correlation are used in the numerator of Equation 4.5. The resulting SAMD values can then be analyzed using the same procedures specified in preceding sections of this chapter.

Computing SAMD Values with Individually Corrected Effect Sizes

Sampling error variances in an individual effect size and in the mean effect size are first computed using uncorrected values as outlined in Equation 4.2 and Equation 4.4. The resulting variances are then divided by the product of the square of the compound attenuation factor and an additional factor denoted by Hunter and Schmidt (1990a, p. 333) as "c." The corrected values of the sampling error variances are then used in the denominator of Equation 4.6 to compute the SAMD statistic while the corrected values of the effect size and the mean effect size are used in the numerator of Equation 4.6. Analysis of resulting SAMD values should then proceed normally when using the procedures specified in preceding sections of this chapter.

The SAS code and data used in the examples presented in this chapter can be downloaded at www.erlbaum.com. Next, chapter 5 presents a summary and guidelines for implementing a meta-analysis.

Summary and Guidelines for Implementing a Meta-Analysis

OVERVIEW

This chapter begins with a summary and overview of the SAS PROC MEANS procedure to meta-analysis and then proceeds to present some general guidelines for implementing a meta-analysis.

SUMMARY OF SAS PROC MEANS PROCEDURE

The methodology for conducting both a Hunter and Schmidt (Hunter & Schmidt, 1990a; Hunter et al., 1982; Schmidt & Hunter, 1977) and a Glassian (Glass, 1976; Glass et al., 1981) meta-analysis using the PROC MEANS procedure in SAS to perform the computations offers several important advantages, including widespread availability both on PCs and mainframes, ease of use, availability of several meta-analysis outcome statistics, and a high degree of flexibility in regard to which corrections can be made for any number and combination of artifacts depending on the information provided in the primary studies. The programs can also be modified to meet the specific needs of the user (e.g., changing the z-values of the confidence and credibility intervals). Furthermore, one particularly attractive feature of this approach is the ease with which moderator variables can be analyzed—by simply adding a few additional statements to the SAS program.

One objective of presenting equations and their associated SAS program code in this volume was to keep researchers and users in touch with the technical aspects of the meta-analysis process. Thus, the step-by-step instructions presented in the preceding chapters should also provide those unfamiliar with the meta-analytic process, a practical means by which they can not only learn, but also actually conduct what has become a useful and common approach to summarizing empirical research (Schmidt, 1992).

There will be, of course, instances or situations where a user might prefer a statistical program other than SAS or even a situation where SAS is not available to the reader. Because the procedures presented in this volume are written strictly for SAS, this may pose a potential limitation. Conversely, given the relatively simple nature of the statistical routines involved (e.g., calculation of means and variances), one possible alternative is to simply translate the procedures to conform to the language of whatever program is available or desired.

GUIDELINES AND POINTERS

This chapter provides a general summary of guidelines and pointers to follow in the implementation of a meta-analysis. Admittedly, although it is a powerful research technique, meta-analyses are not the "be-all, end-all" (Buckley & Russell, 1999; Russell & Gilliland, 1995; Sackett, Tenopyr, Schmitt, & Kehoe, 1985). Meta-analysis is fundamentally a research design and method, and as such, it is only as good as its implementation and execution. Along these lines, it is always good to keep the early "garbage-in, garbage-out" critiques of meta-analysis in mind.

Next, some general guidelines for implementing a meta-analysis are presented. These guidelines incorporate some basic tenets presented by Bobko, Roth, and Potosky (1999) and Rothstein and McDaniel (1989).

Guideline 1

Define the research domain. Have a specific question in mind, and make sure the meta-analytic method chosen can inform that question. Prepare for conducting a meta-analysis by thoroughly familiarizing yourself with the research domain and its major characteristics and idiosyncrasies. Remember that meta-analysis is a research methodology, so unless the study is a methodologically focused meta-analytic study, it is simply a means by which you are trying to investigate or answer specified research questions. Do not let the method drive your research questions; instead, your choice of method should be determined by your research questions. Is meta-analysis the best approach?

Guideline 2

As noted in chapter 1, there are numerous judgements and subjective decisions involved in the implementation of a meta-analysis and these decisions can have profound effects on the meta-analytic results (Wanous et al., 1989). Keep in mind that meta-analysis is not totally objective. It is a more standardized approach to the aggregation of primary studies, compared to a narrative review, but it is not objective. Thus, it is essential that all decisions be clearly stated in the study or research report. Specify the strategy, key words, terms, and phrases used for the electronic search. Also specify the databases searched. Supplement electronic searches with manual search procedures. Given the length of time it takes for journal articles to be published, it may be a good idea to consider conference presentations and other, more current sources. It is important to keep your research questions in mind when determining a time period and databases covered in your search.

Specify the inclusion criteria and decision rules used to select the final set of studies. State the strategy used to extract data from and to code the studies.

Specify the decision to test for moderators, particularly, will this be on a theoretical or conceptual basis (a priori), or will it be empirically determined (post hoc)?

Guideline 3

Look over the data very carefully and do not accept every datum at face value. Ensure that the values coded are accurately extracted. For correlations, decide whether uncorrected or corrected (for artifacts) data points are going to be coded from the primary studies. If corrected correlations are extracted and coded from the primary studies, subsequently correcting them again for artifacts in the meta-analysis would be inappropriate. Exclude errors. Specify the decision to test for and eliminating outliers. Include studies that are methodologically suspect, but only if they are not fatally flawed. Some studies (e.g., Arthur, Bennett, Edens, & Bell, 2001; Burke & Day, 1986; Gaugler, Rosenthal, Thornton, & Bentson, 1987; Kluger & DeNisi, 1996) have coded and tested for the effects of methodological rigor as potential moderators of the results of their meta-analysis.

Guideline 4

For studies with very large sample sizes that may swamp the results, run the meta-analysis with and without these studies and compare the results to determine what effect their absence or presence in the data had on the meta-analysis. The same strategy can be used for outliers and other methodologically suspect studies.

Guideline 5

Specify the decision to aggregate or keep multiple data points separate. Keep in mind that, as a general rule, the preference in meta-analysis is to use independent data points. Thus, if multiple facets of a construct are available, decide whether to average the associated correlations or effect sizes. A rationale should be noted for this decision, especially if the researcher decides to keep them separate. In relation to the issue of independence, ensure that data from primary studies have not been published multiple times, and are not subsequently being used multiple times in the meta-analysis.

Guideline 6

Use multiple coders, check coding reliability, and report the magnitude of interrater agreement. In fact, it is actually a good idea to have an independent person code the studies in addition to the author of the study. The problem with using only the author as a coder is that one risks a loss of objectivity. Because of an author's vested interest in the study, he/she is likely to make biased decisions that may not be made by an independent disinterested coder.

Guideline 7

When making corrections, specify which corrections are made, whether they are individual- or distribution-based corrections. If distribution-based corrections are made, provide information about the artifact distribution.

Guideline 8

When conducting primary research, to facilitate their inclusion in future meta-analyses, make the conscientious effort to collect and report all pertinent information and data. Such information should include, but is not limited to, the pertinent test statistic (e.g., r, t, or F), sample size, means and standard deviations, reliability of the measures, and range restriction.

Guideline 9

Follow the statistical part of the meta-analysis with careful consideration of the results and their possible interpretations. Like other research designs, use meta-analysis as a means of pinpointing the need for additional research. For instance, make a determination whether moderator variables do or do not have a meaningful influence on the strength of the effect.

APPENDIX A

Reference and Information Sources for the Behavioral and Social Sciences

Appendix A presents a listing and description of some reference and information sources (both electronic and text) for the behavioral and social sciences, with slight emphasis on industrial/organization psychology, human resource management, and related disciplines. These sources can typically be found at local university libraries. Most of the research presented here was complied by Tubre, Bly, Edwards, Pritchard, and Simoneaux (2000).

ELECTRONIC DATABASES

General Reference Sources

PsycINFO **(American Psychological Association).** The single most comprehensive reference and information source available to psychologists, *PsycINFO* provides unparalleled access to journal articles, books and book chapters, dissertations, technical reports and other materials from psychology and related disciplines. *PsycINFO* corresponds to the printed volume *Psychological Abstracts* and the CD-ROM product *PsycLIT*, although its coverage is more extensive than either of its companion sources. In June 2000, the database contained more than one million records and had indexed more than 1,400 scholarly journals. Coverage: 1967 to the present (monthly updates).

Social Science Citation Index **(Institute for Scientific Information).** Arguably the most powerful social science literature searching aid currently available, the *SSCI* has indexed 1,725 social science journals and selected social science-related articles from an additional 3,300 journals. The *SSCI* is distinguished by its ability to provide the option of cited reference searching. Using this strategy, a researcher can identify a seminal article in a given area and locate every article published that has since cited the original source. As additional articles are uncovered, the procedure can be repeated in a process called "cycling." Thus, the *SSCI* provides unparalleled access to the progression of a given body of literature. The *SSCI* is published in print, CD-ROM, and online (*Social SciSearch*) formats. Coverage: 1973 to the present (weekly updates for online version).

***ProQuest Digital Dissertations* (formerly *Dissertation Abstracts*, University Microfilms International).** This database contains more than 1.6 million citations to dissertations and selected master's theses covering approximately 3,000 topics. Abstracts are included for records indexed since 1980. This database, available in multiple computer-readable formats (CD-ROM, hard disk, online) corresponds to the print editions of *Dissertation Abstracts International, American Doctoral Dissertations*, and *Comprehensive Dissertation Index*. Coverage: 1861 to the present (monthly updates for online version).

***Current Contents: Social and Behavioral Sciences* (Institute for Scientific Information).** *Current Contents* indexes tables of contents, with complete bibliographic information, for more than 1,375 journals in the social and behavioral sciences. Full length author abstracts are available for approximately 85% of the articles and reviews. One unique and very useful feature is the inclusion of reprint author addresses. *Current Contents* is available in Telnet and Web versions. Coverage: 1969 to the present (rolling 52 weeks; weekly updates).

***PSYNDEXplus With TestFinder* (Zentralstelle fur Psychologische Information and Dokumentation [ZPID] at the University of Trier).** More than 150,000 records on psychology from Germany, Austria, and Switzerland from 1977 to the present in two databases. Compiled by the German Center for Documentation and Information in Psychology at the University of Trier (ZPID), PSYNDEXplus covers German and English journals articles, books, chapters, reports, and dissertations as well as audiovisual media, and in a separate file, approximately 3,700 extensive descriptions can be found of psychological and educational tests used in the German-speaking countries. PSYNDEXplus is bilingual; titles and descriptors are in German and English. German publications are abstracted in German and English publications in English. Test descriptions also include theoretical background, testing procedures, test construction, reliability, validity, fields of application, research and other selected literature, and a comprehensive discussion. PSYNDEXplus is indexed with the Thesaurus of Psychological Index Terms by permission of the American Psychological Association. A bilingual thesaurus is included on disk. In June 2000, the database contained more than 150,000 records and more than 9,500 records are added annually. PSYNDEXplus is available in internet, CD, and hard disk formats. There is no print equivalent of PSYNDEXplus. Coverage: 1977 to the present (quarterly updates).

Sociological Abstracts **(formerly** *SOCIOFILE,* **Sociological Abstracts, Inc.).** *Sociological Abstracts* indexes approximately 2,300 journals covering the literature of sociology and related disciplines, including psychology. Researchers interested in group processes, systems theory, and organizational behavior may find this file of interest. (When it was SOCIOFILE, the user could also access the Social Planning Policy and Development Abstracts (SOPODA) database which covers aspects of applied sociology. *Sociological Abstracts* does not appear to have this feature.) *Sociological Abstracts* is available in multiple computer-readable formats. Coverage: 1963 to the present (monthly updates for online version).

Social Sciences Index **(H. W. Wilson Company).** This database indexes more than 415 periodicals spanning all areas of the social sciences including psychology and sociology. Wilson publishes this index in a variety of formats (CD-ROM, hard disk, online, and print) under the same name. Coverage: 1983 to the present (monthly updates for online version).

Index to Social Sciences and Humanities Proceedings **(Institute for Scientific Information).** *ISSHP* indexes the published proceedings of more than 3,000 conferences held each year. The user can search seven integrated indexes (e.g., author, subject, and meeting location) to locate information from this rich, but often overlooked, resource. *ISSHP* is available in internet, CD-ROM, and print formats. Coverage: 1977 to the present (weekly updates for online version).

Business, Management, Industrial and Organizational Behavior

ABI/INFORM **(University Microfilms International).** This database contains more than 550,000 citations, providing complete bibliographic information, abstracts, and selected full-text articles from more than 1,000 national and international business and management periodicals. *ABI/INFORM's* coverage of human resources management and organizational behavior is particularly relevant. Coverage: 1971 to the present (weekly updates).

Business Periodicals Ondisc is a subset of the *ABI/INFORM* database that provides full-image articles for approximately 400 indexed journals (including *Journal of Applied Psychology* and *Personnel Psychology*). Coverage: 1987 to the present (monthly updates).

Business Periodicals Index **(H. W. Wilson Company).** H. W. Wilson publishes this index to 345 business periodicals in a variety of formats (print, CD-ROM, hard disk, and online) making it widely available. The database covers a broad range of relevant topics including training and business management practices. Coverage: 1982 to the present (monthly updates for online version).

RMNET **(Research Methods Network, Research Methods Division, Academy of Management).** RMNET originated as a question-and-answer discussion group for members of the Research Methods Division (RMD) of the Academy of Management. The RMNET coordinator is Jeff Edwards (jredwards@unc.edu). RMNET has developed into two distinct resources—a question-and-answer network, and an archive on an extensive variety of primarily methodological topics. RMNET has more than 600 members, representing approximately half of the RMD membership. With RMNET, questions can be asked and answers provided to all RMD members who have access to the Internet. A member of RMNET can ask a question by simply sending it to the RMNET e-mail address (rmnet@listserv.unc.edu). Questions sent to RMNET are automatically distributed to all RMNET members who then volunteer answers and send them to the RMNET address. These answers are also automatically distributed to all RMNET members.

With ongoing discussion, the RMNET archive continues to grow in size and value as an important resource for information. The archive can be accessed by e-mail, in which the user sends search phrases to lyris@listserv.unc.edu (e.g., SEARCH RMNET META-ANALYSIS) that then extracts a list of messages that match the search criteria along with the complete text of the messages. The archive can also be searched through the RMNET Web site at http://listserv.unc.edu/cgi-bin/lyris.pl?enter=rmnet&text_mode=0&lang=english.

Access to RMNET is limited to RMD members of the Academy of Management. If you are not a member you will have to ask a colleague or friend who is to do your search for you.

EconLit **(American Economic Association).** Corresponding to the *Journal of Economic Literature* and the *Index of Economic Articles, EconLit* is a comprehensive index to the literature in economics including journal articles, books, dissertations, and conference proceedings. The database provides bibliographic citations, with selected abstracts, and is available in multiple computer-readable formats (CD-ROM, hard disk, and online). Coverage: 1969 to the present (monthly updates for online version).

Testing and Evaluation

ERIC/AE Test Locator **(ERIC Clearinghouse on Assessment and Evaluation).** The *Test Locator* is a joint project of the ERIC Clearinghouse on Assessment and Evaluation, the Library and Reference Services Division of the Educational Testing Service, the Buros Institute for Mental Measurements, the Region III Comprehensive Center, and Pro-ED test publishers. The *Test Locator* consists of the following three files which are accessible, for free, from the ERIC/AE WWW site at the following URL—http://www.ericae.net/testcol.htm.

ETS Test File **(Educational Testing Service).** The *Test Collection* database contains descriptive information including availability, for more than 10,000 tests and research instruments. In addition, users can access *Tests in Microfiche*, an index to approximately 1,000 educational and psychological tests available on microfiche from ETS.

Test Review Locator **(Buros Institute for Mental Measurements; Pro-Ed, Inc.).** This page allows you to search for citations to reviews of educational and psychological tests and measures that The Buros Institute of Mental Measurements and Pro-Ed, Inc. have included in their directories.

Buros/ERIC Test Publisher Directory **(Buros Institute for Mental Measurements).** This page allows you to search for the names and addresses of more than 900 major commercial test publishers.

Education and Educational Psychology

ERIC **(U.S. Department of Education; Educational Resources Information Center).** *ERIC,* sponsored by the U.S. Department of Education, is the single most comprehensive database covering the literature on all aspects of education. The database, available in a variety of electronic formats, corresponds to two print indexes, *Resources in Education (RIE)* and *Current Index to Journals in Education (CIJE).* The *RIE* file covers the document literature in education including such rich sources as conference proceedings and technical reports, while *CIJE* provides comprehensive indexing of nearly 800 periodicals. The *ERIC* database could be an indispensable resource for researchers in the training and development field. Coverage: 1966 to the present (monthly updates for online version and quarterly updates for CD-ROM).

Public Policy, Labor, and Employment Law

LEXIS-NEXIS **(LEXIS-NEXIS, Reed Elsevier, Inc.).** The *LEXIS* service is a premier source of United States federal and state legal and regulatory information. *LEXIS'* comprehensive coverage of state and federal employment and labor law could be particularly useful for industrial/organizational psychology practitioners who seek information concerning legal precedents for issues such as adverse impact, affirmative action, or validity generalization. The *NEXIS* service provides a variety of full-text news, business, and political information. Coverage: Dates of coverage and frequency of updates vary according to source.

PAIS International **(Public Affairs Information Service, Inc.).** *PAIS* is a comprehensive bibliographic index to public and social policy literature in areas including business, economics, and the social sciences. *PAIS'* coverage of emerging trends and legislation in areas such as affirmative action and adverse impact is especially relevant. The database is available in print and electronic formats. Coverage: 1972 to the present (monthly updates for online version).

Index to United States Government Publications

NTIS **(National Technical Information Service).** *NTIS* delivers comprehensive coverage to unclassified technical reports from research sponsored by the U.S. government, foreign governments, federal agencies and other sources. Records include bibliographic information and abstracts with coverage of such relevant areas as management practices, human factors, and public policy. *NTIS* is the sole provider for many of the items located in its database. The database is available in multiple electronic formats and corresponds to the printed source *Government Reports Announcements and Index*. Coverage: 1964 to the present (bi-weekly updates for online version).

PRINTED SOURCES

General Reference Sources

PsycScan: Applied Psychology **(American Psychological Association).** A quarterly publication containing abstracts from a cluster of subscriber-selected applied psychology journals (e.g., *Journal of Applied Psychology, Personnel Psychology, Journal of Occupational and Organizational Psychology, Journal of Management, Academy of Management Journal*, and *Academy of Management Review*). *PsycScan* allows the reader to get an overview of recent research and developments in the field.

Business, Management, Industrial and Organizational Behavior

Human Resources Abstracts **(Sage Publications).** This printed source continues *Poverty and Human Resources Abstracts* (1966–1974). Entries are indexed by author and subject in 15 broad categories. Representative content areas include education and training, hiring and personnel practices, and human resources management. Coverage: 1975 to the present (published quarterly).

Personnel Literature **(U.S. Office of Personnel Management Library).** This government published printed source cites more than 200 journal articles, books, and other documents in each issue under 60-70 broad concept categories that encompasses the entire spectrum of personnel issues. Coverage: 1941 to the present (published monthly).

Personnel Management Abstracts **(Personnel Management Abstracts).** Another printed source, *Personnel Management Abstracts* provides access to articles and books covering a broad spectrum of topics including management development, job design, motivation and many other areas of industrial/organizational psychology. Coverage: 1955 to the present (published quarterly).

Work Related Abstracts **(Harmonie Park Press).** This printed source continues *Employment Relations Abstracts* (1959–1972) and indexes more than 250 serial publications on management and labor. Representative areas of coverage include personnel management and organizational communication. Coverage: 1973 to the present (monthly updates).

Testing and Evaluation

Mental Measurements Yearbook **(Buros Institute for Mental Measurements).** The *Mental Measurements Yearbook (MMY)* series provides descriptive information and critical reviews of commercially available, standardized educational, psychological, and other related English-language tests. Print editions of the *MMY* have been published (in an irregular fashion) since 1938. The thirteenth (and most current) edition was published in 1998. Silver Platter Information provides CD-ROM and online access to later versions of the *MMY*.

Tests in Print **(Buros Institute for Mental Measurements).** *Tests in Print* is a comprehensive bibliographic information source for commercially available English-language tests. Unlike the *MMY, Tests in Print* does not present critical reviews or psychometric information for the tests it covers. The primary value of *Tests in Print* is as a retrospective or master index to the *MMY* series. The latest edition in the irregular publication series is *Tests in Print V* (1999) with 3,009 entries. The first edition of the index was published in 1961.

Test Critiques **(Pro-Ed, Inc.).** Similar in format to the *MMY* series, *Test Critiques* provides descriptive and psychometric information, as well as critical reviews for frequently used psychological, educational, and business-related tests. The most recent edition of *Test Critiques* (Vol. 10) was published in 1994.

Tests **(Pro-Ed, Inc.).** Similar in format to *Tests in Print, Tests* provides descriptive and bibliographic information for a large number of tests used in education, psychology, business and other areas. As with the *MMY-Tests in Print* union, the user must turn to the companion volume, *Test Critiques*, for critical reviews and psychometric information. The fourth and latest edition of *Tests* was published in 1996.

GUIDES, DIRECTORIES, AND OTHER HELPFUL SOURCES

Baxter, P. M. (1993). *Psychology: A guide to reference and information sources.* **Englewood, CO: Libraries Unlimited.** The most recent guide to the literature in psychology, Baxter's book describes nearly 700 reference and information sources available to researchers in psychology and related disciplines. Baxter organizes the work into a number of sections including general psychology reference sources and a number of special topics in psychology. Within each section, a variety of reference and information sources of varying formats (e.g., handbooks and online databases) are identified and discussed.

Books in Print **(R. R. Bowker).** *Books in Print* provides complete, current bibliographic and ordering information for approximately 1.5 million books of all types. Coverage: Current (annual updates).

Douglas, N. E., & Baum, N. (1984). *Library research guide to psychology: Illustrated search strategy and sources.* **Ann Arbor, MI: Pierian Press.** This work is essentially a "how to" manual for library research. A number of basic research tools and their uses are presented by the authors. The volume provides solid advice on developing and carrying out search strategies. However, the dated material of this work may limit its applicability.

Gale Directory of Databases **(Gale Research, Inc.).** This directory provides detailed descriptions of more than 10,000 publicly available electronic databases in a variety of formats. In addition, entries providing contact information and lists of databases are offered for more than 5,000 database producers, vendors, and distributors are included. This database is available in printed and electronic formats. Coverage: Current (semi-annual updates).

Reed, J. G., & Baxter, P. M. (1992). *Library use: A handbook for psychology* **(2nd ed.). Washington, DC: American Psychological Association.** This handbook is an excellent introduction to the process of developing and carrying out a search strategy. Students and professionals who are unfamiliar with basic search and retrieval processes would benefit greatly from the handbook.

APPENDIX B

Equation for Computing the Pooled Within-Group Standard Deviation

The *within-group variance*, when the sample sizes for the experimental and control groups are equal, is defined by (Hunter & Schmidt, 1990a, p. 271) as follows:

$$V_W = \frac{(V_E + V_C)}{2} \tag{B.1}$$

where V_E is the variance for the experimental group, and V_C is the variance for the control group. So, when the sample sizes of the experimental and control groups are equal, the pooled within-group variance is the average of the two within-group variances. Alternately, if the sample sizes are unequal, then a sample size weighted average should be used. Hunter and Schmidt (1990a) weight each variance by its degrees of freedom, that is, by N-1 instead of N. Thus, for unequal sample sizes, the pooled within-group variance is computed as follows:

$$V_w = \frac{[(N_E - 1) * V_E] + [(N_C - 1) * V_C]}{(N_E - 1) + (N_C - 1)} \tag{B.2}$$

where N_E is the sample size for the experimental group and N_C is the sample size for the control group.

The pooled within-group standard deviation is subsequently computed as follows:

$$S_w = \sqrt{V_w} \tag{B.3}$$

APPENDIX C

Conversion and Transformation Equations

COMPUTING ds FOR REPEATED MEASURES OR MATCHED GROUPS DESIGNS

When computing ds for studies with repeated measures or matched groups designs (in contrast to independent groups designs), Dunlap, Cortina, Vaslow, and Burke (1996) demonstrated that the effect size will be overestimated if the test statistic (e.g., t) is converted directly to the effect size (d) without taking the correlation between the pre and post measures into account. Thus, the correct equation for estimating d from t_c (Dunlap et al., 1996, p. 171) follows:

$$d = t_c * \sqrt{\left[\frac{2(1-r)}{n} \right]} \qquad \text{(C.1)}$$

where t_c is a t for correlated observations or repeated measures designs, r is the correlation between measures (i.e., pre and post scores), and n is the sample size. Dunlap et al.'s Appendix A (p. 176) presented the derivation of Equation C.1 for estimating d from t_c.

As a cautionary note, Dunalp et al. concluded that "if the means and standard deviations are not provided, and if the correlation between the measures is not reported nor can be estimated appropriately, then it is best to exclude the study from the meta-analysis rather than risk incorrectly estimating the effect size." (p. 175).

GUIDELINES FOR CONVERTING VARIOUS TEST STATISTICS TO *d* AND *r*

Tables C.1 and C.2 present guidelines for converting various test statistics to *d* and *r*, respectively. The sources for these conversions are Glass et al. (1981), Hunter and Schmidt (1990a), and Wolf (1986).

TABLE C.1
Guidelines for Converting Various Test Statistics to *d*

Statistic to be Converted	Equation for Transformation to d	Comment
t	$$d = \frac{2t}{\sqrt{N}}$$	*N* is the total sample size.
r	$$d = \frac{2r}{\sqrt{1 - r^2}}$$	
F	$$d = \frac{2\sqrt{F}}{\sqrt{df\,(error)}}$$	Use only for comparing two group means (i.e., numerator *df* = 1).
F for experiments using complex analysis of variance		Nouri and Greenberg (1995) presented techniques for estimating effect sizes in the not so uncommon situation where experimental studies use independent variables with more than two levels or analysis of variance with repeated measures. The reader is encourage to consult their article for these techniques.

TABLE C.2
Guidelines for Converting Various Test Statistics to r

Statistic to be Converted	Equation for Transformation to r	Comment
t	$$r = \frac{t}{\sqrt{t^2 + N - 2}}$$	N is the total sample size.
d	$$r = \frac{d}{\sqrt{d^2 + 4}}$$	
F	$$r = \sqrt{\frac{F}{F + df\,(error)}}$$	Use only for comparing two group means (i.e., numerator $df = 1$).
F for $J > 2$ groups.	Collapse J groups to 2 and then convert F to t: $$\lvert t \rvert = \sqrt{F}$$ Or use: $$r = \sqrt{\frac{SS_b}{SS_b + SS_w}}$$	Hays (1973; pp. 683–684).
χ^2 only	$$r = \sqrt{\frac{\chi^2}{N}}$$	N is the sample size. χ^2 only (i.e., no frequencies reported for a contingency table).

(continued)

2 x 2 contingency table	Calculate tetrachoric correlation from tables.	Glass and Stanley (1970; pp. 165 ff)
R x C contingency table	Collapse to a 2 x 2 table and proceed as noted here.	
Point-biserial correlation, r_{pb}	$$r = r_{pb}\sqrt{\frac{n_1 n_2}{u*n}}$$	u = ordinate of unit normal distribution; n = total sample size. Glass and Stanley (1970, p. 171).
Spearman's rank correlation, r_s	The Pearson product–moment correlation is equal to the Spearman's rank correlation because under bivariate normality, the translation of the latter to the former is nearly a straight line.	Kruskal (1958).
Mann-Whitney U	Transform U to r-rank biserial using:	Willson (1976).

$$r_{rb} = \frac{1 - 2U}{(n_1 n_2)}$$

APPENDIX D

Upper Percentage Points of the Chi-square Distribution

TABLE D.1
Upper Percentage Points
of the Chi-Square (χ^2) Distribution

df	.050	.010	.001
1	3.84	6.64	10.83
2	5.99	9.21	13.82
3	7.82	11.34	16.27
4	9.49	13.28	18.47
5	11.07	15.09	20.52
6	12.59	16.81	22.46
7	14.07	18.48	24.32
8	15.51	20.09	26.12
9	16.92	21.67	27.88
10	18.31	23.21	29.59
11	19.68	24.72	31.26
12	21.03	26.22	32.91
13	22.36	27.79	34.53
14	23.68	29.14	36.12
15	25.00	30.58	37.70

(continued)

df	.050	.010	.001
16	26.30	32.00	39.25
17	27.59	33.41	40.79
18	28.87	34.81	42.31
19	30.14	36.19	43.82
20	31.41	37.57	45.32
21	33.67	38.93	46.80
22	33.92	40.29	48.27
23	35.17	41.64	49.73
24	36.42	42.98	51.18
25	37.65	44.31	52.62
26	38.89	45.64	54.05
27	40.11	46.96	55.48
28	41.34	48.28	56.89
29	42.56	49.59	58.30
30	43.77	50.89	59.70
40	55.76	63.69	73.40
50	67.50	76.15	86.66
60	79.08	88.38	99.61
70	90.53	100.40	112.30
80	101.90	112.30	124.80
90	113.10	124.10	137.20
100	124.50	135.80	149.40

References

Abrami, P. C. (1984, February). Using meta-analytic techniques to review the instructional evaluation literature. *Postsecondary Education Newsletter, 6,* 8.

Abrami, P. C., Cohen, P. A., & d'Apollonia, S. (1988). Implementation problems in meta-analysis. *Review of Educational Research, 58,* 151–179.

Alliger, G. M. (1995). The small sample performance of four tests of the difference between pairs of meta-analytically derived effect sizes. *Journal of Management, 21,* 789–799.

Alliger, G. M., Tannenbaum, S. I., Bennett, W. Jr., Traver, H., & Shotland, A. (1997). A meta-analysis of relations among training criteria. *Personnel Psychology, 50,* 341–358.

Aguinis, H., & Pierce, C. A. (1998). Testing moderator variable hypotheses meta-analytically. *Journal of Management, 24,* 577–592.

Arthur, W., Jr., Barrett, G. V., & Alexander, R. A. (1991). Prediction of vehicular accident involvement: A meta-analysis. *Human Performance, 4,* 89–105. An erratum (publisher's correction) to this article was published in *Human Performance, 4,* 231.

Arthur, W., Jr., Bennett, W., Jr., Edens, P. S., & Bell, S. (2001). *Design and evaluation factors that influence the effectiveness of training in organizations: A review and meta-analysis.* Unpublished manuscript. Department of Psychology, Texas A&M University.

Arthur, W. Jr., Bennett, W., & Huffcutt, A. (1994). Choice of software and programs in meta-analysis research: Does it make a difference? *Educational and Psychological Measurement, 54,* 776–787.

Arthur, W., Jr., Bennett, W., Jr., & Huffcutt, A. I. (1995). *Software and programs for conducting meta-analysis research: A Monte Carlo investigation of potential differences* (AL/HR-TR-1995-0092). USAF AMRL Technical Report. Technical Training Division, Brooks Air Force Base, TX.

Arthur, W., Jr., Bennett, W., Jr., Stanush, P. L., & McNelly, T. L. (1998). Factors that influence skill decay and retention: A quantitative review and analysis. *Human Performance, 11,* 57–101.

Arthur, W., Jr., Day, E. A., McNelly, T. L., & Edens, P. S. (2000). *The criterion-related validity of assessment center dimensions: Distinguishing between methods and constructs.* Unpublished manuscript. Psychology Department, Texas A&M University.

Bangert-Drowns, R. L. (1986). Review of developments in meta-analytic method. *Psychological Bulletin, 99,* 388–399.

Barrick, M. R., & Mount, M. K. (1991). The Big-Five personality dimensions in job performance: A meta-analysis. *Personnel Psychology, 44,* 1–26.

167

Bobko, P., Roth, P. L., & Potosky, D. (1999). Derivation and implications of a meta-analytic matrix incorporating cognitive ability, alternative predictors, and job performance. *Personnel Psychology, 52,* 561–589.

Buckley, M. R., & Russell, C. J. (1999). Validity evidence. In R. W. Eder & M. M. Harris (Eds.), *The employment interview handbook* (pp. 35–48). Thousand Oaks, CA: Sage.

Burke, M. J., & Day, R. R. (1986). A cumulative study of the effectiveness of managerial training. *Journal of Applied Psychology, 71,* 232–246.

Callender, J. C., & Osburn, H. G. (1980). Development and test of a new model for validity generalization. *Journal of Applied Psychology, 65,* 543–558.

Carlson, K. D., & Schmidt, F. L. (1999). Impact of experimental design on effect sizes: Findings from the research literature on training. *Journal of Applied Psychology, 84,* 851–862.

Cohen, J. (1988). *Statistical power analysis for the behavioral sciences* (2nd ed.). Hillsdale, NJ: Lawrence Erlbaum Associates.

Cohen, J. (1990). The things I have learned (so far). *American Psychologist, 45,* 1304–1312.

Cohen, J. (1992). A power primer. *Psychological Bulletin, 112,* 155–159.

Cohen, J. (1994). The earth is round ($p < .05$). *American Psychologist, 49,* 997–1003.

Cohen, J., & Cohen, P. (1983). *Applied multiple regression/correlation analysis for the behavioral sciences* (2nd ed.). Hillsdale, NJ: Lawrence Erlbaum Associates.

Cohen, P. A. (1981). Student ratings of instruction and student achievement: A meta-analysis of multisection validity studies. *Review of Educational Research, 51,* 281–309.

Cohen, P. A. (1982). Validity of student ratings in psychology courses: A research synthesis. *Teaching of Psychology, 9,* 78–82.

Cohen, P. A. (1983). Comment on a selective review of the validity of student ratings of teaching. *Journal of Higher Education, 54,* 448–458.

Cohen, P. A. (1986, April). *An updated and expanded meta-analysis of multisection student rating validity studies.* Paper presented at the meeting of the American Educational Research Association, San Francisco.

Cortina, J. M., & Dunlap, W. P (1997). On the logic and purpose of significance testing. *Psychological Methods, 2,* 161–172.

Cortina, J. M., & Gully, S. M. (1999). Ethics of outlier elimination: So the great dragon was cast out . . . who deceives the whole world. *The Academy of Management Research Methods Division Newsletter, 14*(1), 1, 4, 13, 15.

d'Apollonia, S., & Abrami, P. (1996, April). *Variables moderating the validity of student ratings of instruction: A meta-analysis.* Paper presented at the 77th annual meeting of the American Educational Research Association, New York.

d'Apollonia, S., Abrami, P., & Rosenfield, S. (1993, April). *The dimensionality of student ratings of instruction: A meta-analysis of the factor studies.* Paper presented at the annual meeting of the American Educational Research Association, Atlanta, GA.

Dillon, W. R., & Goldstein, M. (1984). *Multivariate analysis.* New York: Wiley.

Dowell, D. A., & Neal, J. A. (1982). A selective review of the validity of student ratings of teaching. *Journal of Higher Education, 53,* 51–62.

Dunlap, W. P., Cortina, J. M., Vaslow, J. B., & Burke, M. J. (1996). Meta-analysis of experiments with matched groups or repeated measures designs. *Psychological Methods, 1,* 170–177.

Freund, R. J., & Littell, R. C. (1991). *SAS system for regression* (2nd ed.). Cary, NC: SAS Institute.

Gaugler, B. B., Rosenthal, D. B., Thornton III, G. C., & Bentson, C. (1987). Meta-analysis of assessment center validity. *Journal of Applied Psychology, 72,* 493–511.

Ghiselli, E. E., Campbell, J. P., & Zedeck, S. (1981). *Measurement theory for the behavioral sciences.* New York: Freeman.

Glass, G. V. (1976). Primary, secondary, meta-analysis research. *Educational Researcher, 5,* 3–8.

Glass , G. V., & Kliegl, R. M. (1983). An apology for research integration in the study of psychotherapy. *Journal of Consulting and Clinical Psychology, 51,* 28–41.

Glass, G. V., McGaw, B., & Smith, M. L. (1981). *Meta-analysis in social science research.* Beverly Hill, CA: Sage.

Glass, G. V., & Stanley, J. C. (1970). *Statistical methods in education and psychology.* Englewood Cliffs, NJ: Prentice-Hall.

Goldberg, L. R. (1992). The development of markers for the Big-Five factor structure. *Psychological Assessment, 4,* 26–42.

Greenfield, T. (1987). Consultants' cameos: A chapter of encounters. In D. J. Hand & B. S. Everitt (Eds.), *The statistical consultant in action.* New York: Cambridge University Press.

Gulliksen, H. (1986). The increasing importance of mathematics in psychological research (Pt. 3). *The Score, 9,* 1–5.

Guzzo, R. A., Jackson, S. E., & Katzell, R. A. (1987). Meta-analysis analysis. *Research in Organizational Behavior, 9,* 407–442.

Hagen, R. L. (1997). In praise of the null hypothesis statistical test. *American Psychologist, 52,* 15–24.

Hays, W. L. (1973). *Statistics for the social sciences.* New York: Holt, Rinehart & Winston.

Hedges, L. V. (1987). How hard is hard science, how soft is soft science: The empirical cumulativeness of research. *American Psychologist, 42,* 443–455.

Hedges, L. V. (1989). An unbiased correction for sampling error in validity generalization studies. *Journal of Applied Psychology, 74,* 469–477.

Hedges, L. V., & Olkin, I. (1985). *Statistical methods for meta-analysis.* Orlando, FL: Academic.

Huber, P. J. (1980). *Robust statistics.* New York: Wiley.

Huffcutt, A. I., & Arthur, W., Jr. (1994). Hunter and Hunter (1984) revisited: Interview validity for entry level jobs. *Journal of Applied Psychology, 79,* 184–190.

Huffcutt, A. I., & Arthur, W., Jr. (1995). Development of a new outlier statistic for meta-analytic data. *Journal of Applied Psychology, 80,* 327–334.

Huffcutt, A. I., Arthur, W. Jr., & Bennett, W. (1993). Conducting meta-analysis using the PROC MEANS procedure in SAS. *Educational and Psychological Measurement, 53,* 119–131.

Huffcutt, A. I., & Roth, P. L. (1998). Racial group differences in employment interview evaluations. *Journal of Applied Psychology, 83,* 179–189.

Huffcutt, A. I., Roth, P. L., & McDaniel, M. A. (1996). A meta-analytic investigation of cognitive ability in employment interview evaluations: Moderating characteristics and implications for incremental validity. *Journal of Applied Psychology, 81,* 459–473.

Huffcutt, A. I., & Woehr, D. J. (1999). Further analysis of employment interview validity: A quantitative evaluation of interviewer-related structuring methods. *Journal of Organizational Behavior, 20,* 549–560.

Hunter, J. E., & Hunter, R. F. (1984). Validity and utility for alternative predictors of job performance. *Psychological Bulletin, 96,* 72–98.

Hunter, J. E., & Schmidt, F. L. (1990a). *Methods of meta-analysis: Correcting error and bias in research findings.* Newbury Park, CA: Sage.

Hunter, J. E., & Schmidt, F. L. (1990b). Dichotomization of continuous variables: The implications for meta-analysis. *Journal of Applied Psychology, 75,* 334–349.

Hunter, J. E., Schmidt, F. L., & Jackson, G. B. (1982). *Meta-analysis: Cumulating research findings across studies.* Beverly Hills, CA: Sage.

Hurtz, G. M., & Donovan, J. J. (2000). Personality and job performance: The big five revisited. *Journal of Applied Psychology, 85,* 869–879.

Iaffaldano, M. T., & Muchinsky, P. M. (1985). Job satisfaction and performance: A meta-analysis. *Psychological Bulletin, 97,* 251–273.

Johnson, B. T. (1989). *DSTAT: Software for the meta-analytic review of research literatures.* Hillsdale, NJ: Lawrence Erlbaum Associates.

Judge, T. A., Thoresen, C. J., Bono, J. E., & Patton, G. K. (2000). *Another look at the relationship between job satisfaction and job performance.* Unpublished manuscript. Department of Management, University of Iowa.

Kittel, J. E. (1957). An experimental study of the effect of external direction during learning on transfer and retention principles. *Journal of Educational Psychology, 48,* 391–405.

Kluger, A. N., & DeNisi, A. (1996). The effect of feedback interventions on performance: A historical review, a meta-analysis, and a preliminary feedback intervention theory. *Psychological Bulletin, 119,* 254–284.

Kruskal, W. H. (1958). Ordinal measures of association. *Journal of the American Statistical Association, 53,* 814–861.

Loehlin, J. C. (1987). *Latent variable models: An introduction to factor, path, and structural analysis.* Hillsdale, NJ: Lawrence Erlbaum Associates.

Lord, R. G., DeVader, C. L., & Alliger, G. M. (1986). A meta-analysis of the relation between personality traits and leadership perceptions: An application of validity generalization procedures. *Journal of Applied Psychology, 71*, 402–410.

Marchese. M. C., & Muchinsky, P. M. (1993). The validity of the employment interview: A meta-analysis. *International Journal of Selection and Assessment, 1*, 18–26.

McCallum, L. W. (1984). A meta-analysis of course evaluation data and its use in the tenure decision. *Research in Higher Education, 21*, 150–158.

McDaniel, M. A. (1986a). Computer programs for calculating meta-analysis statistics. *Educational and Psychological Measurement, 46*, 175–177.

McDaniel, M. A. (1986b). *MAME: Meta-analysis made easy. Computer program and manual, Ver. 2.1.* Bethesda, MD: Author.

McDaniel, M. A., Schmidt, F. L., & Hunter, J. E. (1988). Job experience correlates of job performance. *Journal of Applied Psychology, 73*, 327–330.

McDaniel, M. A, Whetzel, D. L., Schmidt, F. L., & Maurer, S. D. (1994). The validity of employment interviews: A comprehensive review and meta-analysis. *Journal of Applied Psychology, 79*, 599–616.

Mosteller, F., & Hoaglin, D. C. (1991). Preliminary examination of data. In D. C. Hoaglin, R. Mosteller & J. W. Tukey (Eds.), *Fundamental of exploring analysis of variance* (pp. 40–49). New York: Wiley.

Mullen, B. (1989). *Advance BASIC meta-analysis.* Hillsdale, NJ: Lawrence Erlbaum Associates.

Nouri, H., & Greenberg, R. H. (1995). Meta-analytic procedures for estimation of effect sizes in experiments using complex analysis of variance. *Journal of Management, 21*, 801–812.

Ones, D. S., Mount, M. K., Barrick, M. R., & Hunter, J. E. (1994). Personality and job performance: A critique of the Tett, Jackson, and Rothstein (1991) meta-analysis. *Personnel Psychology, 47*, 147–156.

Orr, J. M., Sackett, P. R., & DuBois, C. Z. (1991). Outlier detection and treatment in I/O psychology: A survey of researcher beliefs and an empirical illustration. *Personnel Psychology, 44*, 473–486.

Orwin, R. G., & Cordray, D. S. (1985). Effects of deficient reporting on meta-analysis: A conceptual framework and reanalysis. *Psychological Bulletin, 97*, 134–147.

Petty, M. M., McGee, G. W., & Cavender, J. W. (1984). A meta-analysis of the relationships between individual job satisfaction and individual performance. *Academy of Management Review, 9*, 712–721.

Raju, N. S., & Burke, M. J. (1983). Two new procedures for studying validity generalization. *Journal of Applied Psychology, 68*, 382–395.

Raju, N. S., Burke, M. J., Normand, J., & Langlois, G. M. (1991). A new meta-analytic approach. *Journal of Applied Psychology, 76*, 432–446.

Raven, J. C., Raven, J., & Court, J. H. (1994). *A manual for Raven's Progressive Matrices and Vocabulary Scales.* London: H. K. Lewis.

Rosenthal, R. (1978). Combining results of independent studies. *Psychological Bulletin, 85,* 185–193.

Rosenthal, R. (1983). Assessing the statistical and social importance of the effects of psychotherapy. *Journal of Consulting and Clinical Psychology, 51,* 4–13.

Rosenthal, R. (1984). *Meta-analytic procedures for social research.* Beverly Hills, CA: Sage.

Rosenthal, R., & Rubin, D. B. (1982). Comparing effect sizes of independent studies. *Psychological Bulletin, 92,* 500–504.

Rosenthal, R., & Rubin, D. B. (1986). Meta-analytic procedures for comparing studies with multiple effect sizes. *Psychological Bulletin, 99,* 400–406.

Rosnow, R. L., Rosenthal, R., & Rubin, D. B. (2000). Contrasts and correlations in effect-size estimation. *Psychological Science, 11,* 446–453.

Rothstein, H. R., & McDaniel, M. A. (1989). Guidelines for conducting and reporting meta-analyses. *Psychological Reports, 65,* 759–770.

Russell, C. J., & Gilliland, S. W. (1995). Why meta-analysis doesn't tell us what the data really mean: Distinguishing between moderator effects and moderator processes. *Journal of Management, 21,* 813–831.

SAS Institute, Inc. (1990). *SAS Language: Reference, Version 6* (1st ed.). Cary, NC: Author.

Sackett, P. R., Tenopyr, M. L., Schmitt, N., & Kehoe, J. (1985). Commentary on forty questions about validity generalizations and meta-analysis. *Personnel Psychology, 38,* 697–798.

Scarpello, V., & Campbell, J. P. (1983). Job satisfaction: Are all the parts there? *Personnel Psychology, 36,* 577–600.

Schmidt, F. L. (1992). What do data really mean? *American Psychologist, 47,* 1173–1181.

Schmidt, F. L. (1996). Statistical significance testing and cumulative knowledge in psychology: Implications for the training of researchers. *Psychological Methods, 1,* 115–129.

Schmidt, F. L., & Hunter, J. E. (1977). Development of a general solution to the problem of validity generalization. *Journal of Applied Psychology, 62,* 529–540.

Schmidt, F. L., & Hunter, J. E. (1978). Moderator research and the law of small numbers. *Personnel Psychology, 31,* 215–232.

Schmidt, F. L, & Hunter, J. E. (1998). The validity and utility of selection methods in personnel psychology: Practical and theoretical implications of 85 years of research findings. *Psychological Bulletin, 124,* 262–274.

Schmidt, F. L., Hunter, J. E., Pearlman, K., & Hirsh, H. R. (1985). Forty questions about validity generalization and meta-analysis. *Personnel Psychology, 38,* 697–798.

Schmidt, F. L., Hunter, J. E., & Raju, N. S. (1988). Validity generalization and situational specificity: A second look at the 75% rule and Fisher's z transformation. *Journal of Applied Psychology, 73,* 665–672.

Schmidt, F. L., Law, K., Hunter, J. E., Rothstein, H. R., Pearlman, K., & McDaniel, M. (1993). Refinements in validity generalization methods: Implications for the situation specificity hypothesis. *Journal of Applied Psychology, 78,* 3–12.

Schmitt, N., Gooding, R. Z, Noe, R. A., & Kirsch, M. (1984). Metaanalyses of validity studies published between 1964 and 1982 and the investigation of study characteristics. *Personnel Psychology, 37,* 407–422.

Shuell, T. J., & Keppel, G. (1970). Learning ability and retention. *Journal of Educational Psychology, 61,* 59–65.

Stajkovic, A. D., & Luthans, F. (1998). Self-efficacy and work-related performance: A Meta-analysis. *Psychological Bulletin, 124,* 240–261.

Stauffer, J. (1994). *MetaDOS—A PC-based meta-analysis program.* Terre Haute: Indiana State University.

Steiner, D. D., Lane, I. M., Dobbins, G. H., Schnur, A., & McConnell, S. (1991). A review of meta-analyses in organizational behavior and human resources management: An empirical assessment. *Educational and Psychological Measurement, 51,* 609–626.

Tett, R. P., Jackson, D. N., & Rothstein, M. (1991). Personality measures as predictors of job performance: A meta-analytic review. *Personnel Psychology, 44,* 703–742.

Tett, R. P., Jackson, D. N., Rothstein, M., & Reddon, J. R. (1994). Meta-analysis of personality-job performance relations: A reply to Ones, Mount, Barrick, and Hunter (1994). *Personnel Psychology, 47,* 157–172.

Therrien, M. E. (1979, September). Evaluating empathy skill training for parents. *Social Work,* 417–419.

Thompson, B. (1996). AERA editorial policies regarding statistical significance testing: Three suggested reforms. *Educational Researcher, 25,* 26–30.

Tubre, T. C., Bly, P. R., Edwards, B. D., Pritchard, R. D., & Simoneaux, S. (2000). *Building a better literature review: Reference and information sources for industrial/organizational psychology.* Manuscript under preparation. College Station: Texas A&M University.

Tukey, J. W. (1960). A survey of sampling from contaminated distributions. In I. Olkin, J. G. Ghurye, W. Hoeffding, W. G. Madoo & H. B. Mann (Eds.), *Contributions to probability and statistics* (pp. 448–485). Stanford, CA: Stanford University Press.

Tukey, J. W. (1977). *Exploratory data analysis.* Reading, MA: Addison-Wesley.

Wanous, J. P., & Reichers, A. E. (1996). Estimating the reliability of a single item measure. *Psychological Reports, 78,* 631–634.

Wanous, J. P., Reichers, A. E., & Hudy, M. J. (1997). Overall job satisfaction: How good are single-item measures? *Journal of Applied Psychology, 82,* 247–252.

Wanous, J. P., Sullivan, S. H., & Malinak, J. (1989). The role of judgement calls in meta-analysis. *Journal of Applied Psychology, 74,* 259–264.

Wesman, A. G. (1965). *Manual for Wesman Personnel Classification Test.* New York: The Psychological Corporation.

Whitener, E. M. (1990). Confusion of confidence intervals and credibility intervals in meta-analysis. *Journal of Applied Psychology, 75*, 315–321.

Wiesner, W. H., & Cronshaw, S. F. (1988). A meta-analytic investigation of the impact of interview format and degree of structure on the validity of the employment interview. *Journal of Occupational Psychology, 61*, 275–290.

Wilkinson, L., and the Task Force on Statistical Inference. (1999). Statistical methods in psychology journals: Guidelines and explanations. *American Psychologist, 54*, 594–604.

Willson, V. L. (1976). Critical values of the rank-biserial correlation coefficient. *Educational and Psychological Measurement, 36*, 297–300.

Wolf, F. M. (1986). *Meta-analysis: Quantitative methods for research synthesis*. Beverly Hills, CA: Sage.

Author Index

175

Subject Index

Milton Keynes UK
Ingram Content Group UK Ltd.
UKHW022107141024
449569UK00031B/1818